Recent Advances in China-Australia Managed Aquifer Recharge

中国–澳大利亚含水层补给管理新进展

主编　王维平　Peter Dillon　Joanne Vanderzalm

黄 河 水 利 出 版 社

·郑州·

内 容 提 要

本书精选了中国–澳大利亚含水层补给管理研讨会论文，其主要内容包括可管理含水层补给的地址的选择、设计，有关水力学知识，地球化学特征及水质变化，风险评估，数学模型及模拟、指南，中国和澳大利亚有关可管理含水层补给的案例研究等。

本书适合从事地下水科研、教学、管理及相关部门人员参考使用。

图书在版编目(CIP)数据

中国–澳大利亚含水层补给管理新进展/王维平，(澳)迪伦(Dillon, P.)，(澳)范德赞姆(Vanderzalm, J.)主编. —郑州：黄河水利出版社，2009.3

ISBN 978-7-80734-572-5

Ⅰ.中… Ⅱ.①王… ②迪… ③范… Ⅲ.含水层–补给水–水资源管理–研究–中国、澳大利亚 Ⅳ.P641

中国版本图书馆 CIP 数据核字(2009)第 026174 号

组稿编辑：王路平　电话：0371-66022212　E-mail：hhslwlp@126.com

出 版 社：黄河水利出版社

地址：河南省郑州市顺河路黄委会综合楼 14 层　邮政编码：450003

发行单位：黄河水利出版社

发行部电话：0371–66026940、66020550、66028024、66022620(传真)

E-mail：hhslcbs@126.com

承印单位：河南省瑞光印务股份有限公司

开本：787 mm × 1 092 mm　1 / 16

印张：13.5

字数：310 千字　　印数：1—1 000

版次：2009 年 3 月第 1 版　　印次：2009 年 3 月第 1 次印刷

定价：35.00 元

《中国–澳大利亚含水层补给管理新进展》

编 委 会

Preface

Managed aquifer recharge (MAR) is the intentional recharge of water to aquifers for subsequent recovery or environmental benefit. Aquifers may be recharged by a range of methods using wells and infiltration systems, to increase groundwater supplies and improve their quality or to redress saline ingress or land subsidence.

To date in China the major applications of MAR have been in infiltration basins established in the 1990s, such as the groundwater dams, infiltration wells and trenches in Shandong province, with a total recharge capacity exceeding 200 million m^3/a. River bank filtration projects have been established in South China to improve the quality of water for industry and agriculture. Injection wells are now in routine operational use for sustainable management of geothermal reservoirs in North East China with reinjection of 4 million m^3/a. In addition, there are recharge projects to reduce land subsidence in Shanghai. A demonstration project in Beijing is exploring wastewater treatment processes to precede recharge for sustainable projects that protect human health. New research in Jinan is starting to explore recharge of roof rainwater to replenish a karst aquifer that feeds historically important springs.

In Australia infiltration basins have been used, for 50 million m^3/a of agricultural supplies, since the late 1970s, commencing in the Burdekin Delta, Queensland. More recently urban stormwater and reclaimed water resources have been used in MAR, via injection wells in South Australia and infiltration projects utilising soil aquifer treatment in Western Australia and Northern Territory.

The China-Australia Managed Aquifer Recharge (MAR) Training Workshop held in Jinan October 27 ~ 31, 2008, was supported by the AusAID Public Sector Linkages Program, the National Natural Science Foundation of China, Shandong Institute of Geological Survey and Jinan Water Resources Bureau in order to facilitate the exchange of information and experience on MAR among Australian researchers and Chinese researchers and water resource managers. In large parts of both countries groundwater is a resource under stress, and in urban areas the possibilities of urban runoff and treated sewage for augmenting groundwater supplies are being trialled or used. This workshop proceedings documents the way MAR is being used or could be developed to address some of China's water resource management challenges. It provides a guide to developing MAR schemes based on Australian and Chinese experience. The material presented takes account of technical, health or environmental risks and the methods used to manage those risks. Consequently MAR operations can be confidently sited, designed, constructed and operated to be successful for a range of purposes.

As an outcome of the workshop research collaborations are intended to emerge among Chinese researchers, and between them and Australian counterparts. Progress on developing and communicating the outcomes of demonstration projects, providing unified national guidance and policies for MAR, and developing maps of the potential opportunities for MAR are recommendations emerging from this workshop. In addition, and to support this progress, a MAR Network was formed to facilitate ongoing communication between researchers and exponents of MAR. Dr Wang Weiping and University of Jinan will provide the hub of this network as part of an international UNESCO network. Workshop attendees hope this will spread quickly throughout China where MAR has a valuable role to play in water supply development and protection particularly where access to safe drinking water supplies is limited.

Peter Dillon　Wang Weiping

January 2009

前　言

中国北方地区水资源缺乏，供需矛盾尖锐，地下水严重超采，并带来一系列生态环境问题，例如大面积地下水漏斗、地面沉降、岩溶塌陷、泉水停喷、海水入侵以及地下水水质恶化等。可管理的含水层补给技术采用雨洪水或者再生水人工回灌地下水，将含水层作为地下库容，既可在时间和空间上调节水资源、增加水资源可利用量、改善水质，又可以预防和修复由于地下水超采带来的生态环境问题，是保证城市用水安全的最经济的方法，对北方缺水城市具有重大的经济效益、社会效益和生态环境效益。

澳大利亚联邦科学与工业研究组织(CSIRO)、清华大学、济南大学、中国科学院广州地球化学研究所、中国地质调查局于2008年10月27～31日在济南大学主办了中国–澳大利亚含水层补给管理研讨会。国际水文地质学家协会含水层补给管理专业委员会主席Peter Dillon博士，出席了研讨会，并进行了讲座。经过十多年的研究和推广，目前澳大利亚MAR(Managed Aquifer Recharge)技术每年供水300万m^3，计划每年利用城市雨水和再生水回灌含水层开发25 000万m^3补充供水。研讨会的目的是将澳大利亚在含水层补给管理方面的技术和经验介绍到中国，促进MAR在中国的发展；同时交流中国在人工回灌地下水的案例和经验。

研讨会上中澳双方的专家和教授从不同的方面，多角度地对人工回灌地下水的有关问题进行了讲座和研讨，澳方研讨的内容涉及MAR地址的选择、设计和建设，水力学知识，地球化学特征及水质的变化，风险评估、指南与模拟和澳大利亚有关可管理含水层补给的案例；中方研讨的内容有地热水回灌深层含水层、城市污水处理后的再生水回灌浅层孔隙含水层、城市屋顶雨水回灌岩溶含水层、滨海地区建地下坝拦蓄地下径流并利用汛期雨洪水、地表水回灌浅层含水层等，并组织编辑成论文集。

研讨会得到了澳大利亚政府澳大利亚国际开发署(Australian Government AusAID)、澳大利亚联邦科学与工业研究组织(CSIRO)、中国国家自然科学基金委员会(National Natural Science Foundation of China(40810162))、济南大学、济南市水利局、山东省地质调查院资助。

本书的出版得到了中国国家自然科学基金委员会(National Natural Science Foundation of China(40742014、40810162))、山东省重点学科(实验室)基金项目、山东省自然科学基金(Y2008E18)、济南市科技发展计划(企业自主创新计划200807125)项目、澳大利亚联邦科学与工业研究组织(CSIRO)等的资助。

书中不当之处，敬请读者批评指正。

作　者

2008年12月

前言

目　录

目　录

Preface

特约论文

Identification of [illegible]

[illegible]

Study on [illegible]

[illegible]

[illegible] with [illegible]

[illegible]

Introduction to Integrated Water Resources [illegible]

[illegible]

Water Quality [illegible]

[illegible]

Reinjection of Geothermal Water in Beijing and Tianjin Areas of China

Han Zaisheng[1], Liu Jiurong[2], Wang Kun[3]

(1 China Geological Survey, No.45 Fuwai Street, Beijing, 100037; 2 Beijing Institute of Geological Engineering, No.8 Beiwa Road, Balizhuang, Haidian District, Beijing, 100037; 3 Tianjin Bureau of Land Resources and Real Estate Management, No.84 Qufu Road, Heping District, Tianjin, 300042)

Abstract: Beijing and Tianjin are two very big cities with abundant low-temperature geothermal in China, using it mostly for space heating, bathing, health spas and recreation, fish farming, greenhouses and domestic hot water etc. With the improving living conditions of the citizens, the need for geothermal water is ever increasing in the cities. Large-scale abstraction of geothermal water caused a rapid declining of the water level of the geothermal wells in both Beijing and Tianjin, and how to ensure the sustainability of geothermal utilization has been taken as the most important task. Apart from the limitation of geothermal water production, reinjection is also taken as an important aspect of geothermal management. Reinjection has been widely applied in the management of geothermal resources, and is becoming a kind of routine in a lot of geothermal fields. In China, reinjection tests started in 1974 in the City of Beijing. But reinjection of used geothermal water from heating systems in scale was started in the City of Tianjin in the 1990's. Since 2000, reinjection has been taken as a more and more important aspect of geothermal resources management in the two cities. At present, a number of reinjection projects have been in operation in Beijing and Tianjin. As a result, the water level declining of the geothermal wells has been greatly mitigated. In Xiaotangshan geothermal field of Beijing, the water level has even stopped declining since 2006, owing to the reinjection and controlling of the production of geothermal water. The reinjection history in both Beijing and Tianjin will be summarized, and the experiences and problems of reinjection, including the distance between the production and reinjection wells, tracer test and monitoring of geothermal fields with reinjection will be discussed.

Keyword: Geothermal reinjection; management

1 Introduction

Geothermal is a kind of green energy, widely used in a lot of countries for power generation and direct purpose, such as space heating, bathing, swimming pools, fish farming, greenhouses, health spas and recreation etc, creating significant economical and environmental effects. Geothermal is a kind of renewable energy, but it should not be over-exploited; otherwise, the resources will be depleted, or will need a rather long time to recover from the improper management. Therefore, it is mandatory to implement proper management for the sustainable use of geothermal resources. Reinjection has been widely used in the management of geothermal fields, and is becoming a kind of routine in a lot of geothermal fields, since the first such project was implemented in the famous Geysers in 1969 (Axelsson and Stefansson, 1999). The purpose of geothermal reinjection is for ①the disposal of the waste geothermal fluid that may cause thermal and chemical pollution to the environment; ②the improvement of the heat mining, because over 90% of the heat in the geothermal reservoirs is stored in the hot rock matrix; ③the stabilization of the production capacity of the geothermal field through the maintenance of the reservoir pressure (Liu, 1999). Geothermal reinjection started out as a method of disposing of wastewater from power plants in order to protect the surrounding environment. It started as early as 1969 and 1970 at the Geysers in California and Ahuachapan field in El Salvador, respectively. Presently there are a number of geothermal fields worldwide where injection is already a part of the field operation, including the Geysers field in USA, Larderrello field in Italy, Berlin field in El Salvador, Paris in France, Laugaland field in Iceland etc. There are a number of other geothermal fields where reinjection experiments have been carried out, and some of them may start production-scale reinjection soon. There are abundant low enthalpy geothermal resources in China (high enthalpy geothermal only exists in Tibet). It is mostly used for health spas and recreation in southern China because it is not very cold even in the winters; and it is used for various direct purposes in northern China where it is very cold in the winters. For over 30 years, geothermal utilization has been ever increasing, especially in the past 10 years and in some northern big cities such as Beijing, Tianjin and Xi'an etc (Han, 2002). With the expansion of geothermal utilization, some problems have been found, for example, the rapid declining of the reservoir pressure in the geothermal fields where the production is of a large-scale. Therefore, reinjection is considered as a kind of measure for the sustainable use of geothermal energy. In China, the earliest geothermal reinjection experiments were started in the urban area of Beijing in 1974 and 1975. In 1980, larger-scale reinjection experiments were carried out in the geothermal area: cold groundwater and return geothermal water was injected into a geothermal well as deep as 1 275 m. At the end of the 1980's, reinjection tests were carried out in the Tertiary sandstone reservoir in Tianjin. Since 1996, reinjection experiments have been implemented in the dolomite reservoir in Tianjin. Till now, there have been 13 production-reinjection doublets running in Tianjin. In 2004 and 2005, reinjection experiments into the sandstone reservoir were carried out in Tianjin again. In 2001, reinjection experiments were implemented in the Xiaotangshan geothermal field north of Beijing. Since then, production scale reinjection started in Xiaotangshan Geothermal field. Experiments in both Tianjin and Beijing showed that reinjection is a feasible measure to ensure the sustainable use of geothermal resources in the two cities.

2 The importance of geothermal reinjection

Beijing is rich with low-temperature geothermal stored in limestone or dolomite reservoirs, and the areas that have been identified with geothermal potential are over 2 760 km^2, divided into 10 geothermal fields, such as the ones in the southeast urban area and Xiaotangshan (about 30 km north of the city centre). The temperature of geothermal water in Beijing is 38 ~ 89 °C. The geothermal water contains SiO_2 and other components that are good for human health. In the history, hot spring water was used for bathing and spas in Beijing. Large-scale geothermal use only started in 1971 in Beijing, with the completion of the first geothermal well. After that, the number of geothermal wells increased very fast, and the amount of geothermal water production increased in the mean time. By 1985, the geothermal production increased to over 10×10^6 m^3 annually, causing a rapid declining of reservoir pressure (water level). Therefore, strict measures were taken to control the amount of geothermal water abstraction since 1985. As a result, the water level decline has slowed down since then. At present, it is often 1~2.5 m annually, and still threatens the sustainability of geothermal utilization in Beijing.

This means that the net discharge of geothermal water, that is, the gap between production and reinjection, of the geothermal systems should be less than a certain limit. Otherwise, the water level of the geothermal wells will decline, and threatens the sustainability of geothermal use. It also means that if reinjection is not carried out, the production will have to be decreased by a large amount. Therefore, it is essential to reinject the used geothermal water back to the geothermal reservoir properly. The geothermal resources in Tianjin are of a typical low-enthalpy geothermal in sedimentary basin. The area with geothermal potential is about 8 700 km^2, accounting for about 77% of the total area of the city. Geothermal is stored in Tertiary sandstone and the karst/fractured dolomite reservoirs. The temperature of the geothermal water ranges from 55 °C to103 °C. Geothermal is widely used for space heating, domestic hot water, fish farming and greenhouses, and recreation etc. By the end of 2007, 294 geothermal wells (including 38 reinjection wells) have been drilled in Tianjin, of which the deepest is close to 4 000 m. The production capacity of each well is 100~200 m^3/h. The annual production of geothermal water was 24.50×10^6 m^3.

Due to the large-scale development of the geothermal resources, the reservoir pressure decreases quickly in Tianjin, especially in the dolomite reservoir. Since 1997, the annual water level drawdown has been over 3 m. Currently, the depth to the static water level in the geothermal wells varies between –40 m and –90 m, with an annual drawdown of 6 ~ 9 m (see Figure1). This suggests that the recharge to the reservoirs is rather limited. Therefore, it is necessary to implement reinjection for maintaining the reservoir pressure and prolonging the life time of the geothermal wells.

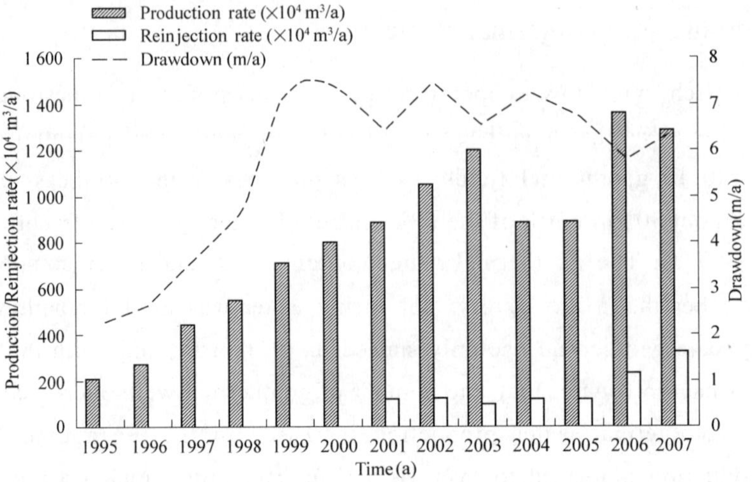

Figure 1　The history curve of water level drawdown vs geothermal water production and reinjection of the dolomite reservoir in the Tianjin urban area in 1995 ~ 2007

3　Reinjection of geothermal water in Beijing

3.1　History

In China, the earliest geothermal reinjection experiments were started in the urban area of Beijing in 1974 and 1975. In 1980, larger-scale reinjection experiments were carried out in the geothermal area: cold groundwater and return geothermal water was injected into a geothermal well as deep as 1 275 m. After then, reinjection stopped in Beijing for a rather long time. Until 2001, reinjection experiment and demonstration projects started again. Recently, reinjection has been implemented in rather large scales and got significant achievements in Beijing. The geothermal reinjection history can be divided into 4 periods:

(1)1974~1983. Because of the increase of geothermal water production in early 1970's, the water level of the geothermal wells declined rather fast. It was considered that reinjection would be an important measure to prolong the lifetime of the geothermal wells. In 1974, a short reinjection test was conducted in the famous Park of Temple of Heaven in the southeast urban area. And in 1975, a short reinjection test was conducted again in the southeast urban area. To 1980, there have been over 30 geothermal wells in the southeast part of Beijing, and the geothermal water production has been about 3 million m^3/a, and the water level decline became more serious. Therefore, a larger-scale injection test was carried out from June 4 to September 2, 1980. Cold groundwater of 15.5 ℃ was injected into a 1 060 m deep geothermal well in the southeast urban area. In the 89 d of injection test, about 60 000 m^3 of cold water was injected, and the temperature change inside the injection well was observed (Liu *et al.*, 1981). After that, 40 ℃ return geothermal water from a heating system was reinjected into a 1 274.65 m deep geothermal well in the same area (Bai and Gong, 1984). In that period, reinjection tests focused on after how much time the injected colder water could be heated again in the same wells.

(2)1983 ~ 2000. Geothermal reinjection activity stopped mostly because of financial problems, although the water level continued to decline in this period (Han, 2002).

(3)2001 ~ 2002. After 1999, with the fast development of geothermal utilization in Beijing, and a wide acceptance of the idea of sustainable development, reinjection was considered again to counter the decline of the water level. In 2001, a demonstration project of geothermal reinjection was carried out in the Xiaotangshan area. Return water from geothermal heating was injected back to the same geothermal reservoir through a well about 200 m away from the production well. This reinjection project operated smoothly in 2001 and 2002, and did not show any cooling of the production temperature (Liu, 2003). In 2001, a reinjection test was also conducted in the southeast urban area in Beijing. Return water from a heating system was injected into a well that struck a shallower geothermal reservoir through a well 80 m away from the production well. And reinjection experiments also started in 2002 in the southeast boarder of the same geothermal field (Liu, 2003).

(4)2003 to present. Since 2003, reinjection in Beijing expanded rapidly, after the experiments in 2001 and 2002. In the Xiaotangshan area, there have been 6 reinjection wells in operation in 2006, and the amount of water reinjection has been about 60% of the production. Reinjection also expanded in other parts of Beijing in this period, and the role of reinjection on sustainable utilization of geothermal resources has been well presented.

3.2 Reinjection in the Xiaotangshan area

The reinjection in Xiaotangshan was started in 2001 in a hotel in the centre part of the geothermal field. One of the two production wells of the hotel was converted into an injection well. The wells of the hotel were drilled in 1984 and 1996, respectively. The distance between the wells is about 200 m. The geothermal reservoir is in the limestone of Cambrian System and the Dolomite of the Jixian System. The wells encountered the same fault that is very important to the occurrence of geothermal around the area of the hotel. The reinjection was carried out from November 30, 2001 to March 27, 2002, totaling 117 d. The temperature of the reinjected return geothermal water was 30 ~ 44 °C. The flow rate of reinjection changed with the atmospheric temperature. It was around 800 m^3/d in the coldest days from 8 to 20, January, 2002, and was under 800 m^3/d on the rest of the days. The injectivity of the well did not decrease during the injection. The total amount of water injected was 73 331 m^3 in that heating season (Liu and Yan, 2006).

A tracer test was conducted during the reinjection. On January 8, 2002, 50 kg KI was applied to the reinjection well instantaneously, 39 d after the injection started. 165 water samples from the production well were collected till the space heating stopped. Some samples were also collected from the surrounding wells. But there was not any iodine found in the samples. This indicates that there is not a direct pass between the reinjection and production wells, and premature thermal breakthrough is not likely to happen in the production well (Liu, 2002). After the injection stopped, a submersible pump was installed in the injection well, intended to restore the injectivity of the well, if there was any reduction. On April 15, 2002, the pump was started. At the beginning, the temperature of the water was around 30 °C, and in an hour, the water became 63.5 °C, nearly restored to its normal production temperature that is 64 °C.The reinjection experiment shows that the injectivity of the geothermal reservoir is rather good, and the reservoir also has a good capacity to heat the reinjected colder water (Liu, 2006).

In the 2003 ~ 2004 heating period, a reinjection test was carried out in another hotel very close to the first one. In 150 d, 1.48×10^5 m^3 of return geothermal water from the heating system was injected into a well drilled for injection purpose. The total amount of reinjection reached 2.48×10^5 m^3 in that heating season. In the 2004 ~ 2005 heating period, 4 production-reinjection doublet systems were set up, by converting old production wells into injection wells, or by drilling new injection wells in the geothermal field. From November, 2004, to April, 2005, 10.2×10^5 m^3 of return geothermal water was injected into the geothermal reservoir, accounting for 36.5% of the total production (Liu and Yan, 2006). In the 2005~2006 heating period, there have been 6 reinjection wells for injecting the return water of 8 production wells. There are 2 production-reinjection assemblages that each involves 2 production wells and 1 reinjection well. The total quantity of reinjection was 1 322 778 m^3, accounting for 56.6% of the annual production in the field.

In the 2001 ~ 2002, the effect of the reinjection on the stabilization of the reservoir pressure was very little, because the amount of reinjection was little. With the increase of reinjection, the effect became more and more significant. In the 5 months from December, 2004, to April, 2005, the water level of the monitoring well (has been monitored for about 30 years) was higher than that in the same period in 2003 and 2004(See Figure2), it rose 2.5 m. Considering that the water level decreased 1 ~ 1.5 m every year before the large-scale reinjection, the effect was very significant.

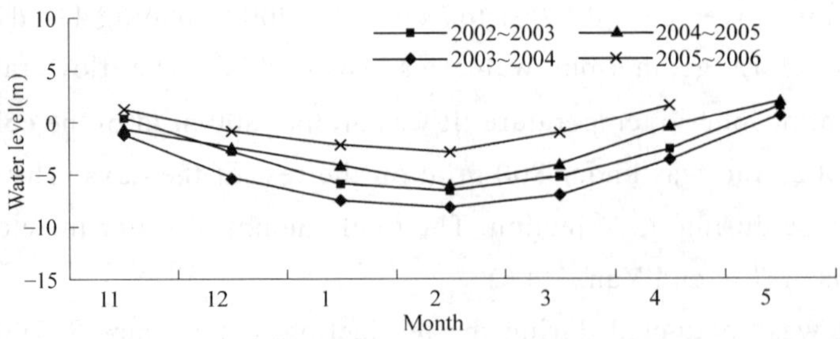

Figure 2　The water level in Xiaotangshan geothermal field from 2002 to 2006

The reinjection in the Xiaotangshan geothermal field does not have observable influence on the temperature of production wells, although the distance between some of the production and reinjection wells is as short as 200 m. It has also been found out that the chemical composition of the geothermal water from the production wells did not change apparently. But the content of HCO_3^- of the water pumped from one of the reinjection wells decreased, and the content of SO_4^{2-} increased. This may indicate that the reinjected colder water flows to the deeper part of the geothermal reservoir and the hotter water from a greater depth flows to the top of the reservoir.

3.3 Reinjection in other areas in Beijing

Although geothermal reinjection tests have been carried out as early as 1974 in the urban area in Beijing, long term reinjection only started in early 2002 in an apartment building district about 5 km south of the Tiananmen Square. There are two geothermal reservoirs at different depths, both made of dolomite, and separated by a shale layer about 100 m thick. Two wells, 90 m apart from each other, were drilled in 2001 for the space heating of the 28 000 m^2 floor area (with the help of a heat pump system) and the reinjection of the tail water from the heating system. The reinjection well is 1 900 m deep, striking the upper reservoir; the production well is 2 054 m deep, completed for producing from the lower reservoir. The water temperature from the reinjection well and the production well is 54 °C and 59 °C respectively. Both wells have a rather good production capacity. The average flow rate of geothermal water in the heating system was 35 m^3/h, and all of the tail water was reinjected into the upper reservoir. The water level rose approximately 4 m on average in the injection well. The experiment shows that the injectivity of the well is close to its productivity. This geothermal heating system, incorporating reinjection and heat pumps, has been running for more than 6 years, and did not meet any difficulties. It is also a good example for the cascaded use of geothermal resources (Liu, 2006). After that, a few other reinjection wells were put into use in the southeast urban area and other geothermal fields in Beijing. Because the government encourages reinjection by deducting a geothermal resources fee, more and more users are planning to start reinjection.

4 Reinjection of geothermal water in Tianjin

4.1 History

Reinjection activity started in the early 1980's in Tianjin, and the history may be divided into 3 periods:

(1)1980~1995.With the increase of geothermal water production, it was realized that reinjection might be useful to control the declining water level. At the beginning of the 1980's, studies on geothermal production-reinjection doublets or production-reinjection well groups were carried out, and related numerical modeling on sandstone reinjection was conducted. According to the results, reinjection tests were carried out in Dagang District and Tanggu District in 1987~1989 and 1995, focused on reinjection in the sandstone geothermal reservoir of the Tertiary System.

(2)1995~1998. With the increase of geothermal water use from the dolomite reservoir, more attention was paid on reinjection into the dolomite reservoir in Tianjin. In 1995, the first such production-reinjection doublet was drilled, and a reinjection test was carried out in 1996~1997. The result showed that reinjection is technically feasible in the dolomite geothermal reservoir in Tianjin (Wang, *et al.*, 2001).

(3)1998 to present. Based on the results of the reinjection tests for the dolomite reservoir, a few production and reinjection doublets were installed and operated in this period in Tianjin. In the same time, tracer tests were carried out for understanding the movement and heating process of the reinjected colder water. Monitoring of the doublets has become routine, including the flow rate, water level, temperature, chemical contents. And numerical modeling was carried out for predicting the effect of reinjection. A technical standard about the design and operation of geothermal reinjection was compiled, by summarizing the experiences of geothermal reinjection in Tianjin. Since 2004, a test on reinjection into the sandstone geothermal reservoir was started again in Tianjin, due to the rapid decline of the geothermal water level in the sandstone reservoir.

4.2 Dolomite reservoir reinjection

Most of the geothermal production-reinjection doublet systems in Tianjin are inside the urban area. Both of the production and reinjection wells were drilled into the dolomite reservoir which spreads widely in the area of Tianjin (Wang *et al.*, 2001).

Since the first geothermal production-reinjection doublet was put into operation in the winter (during the space-heating period) in 1999, there have been 27 reinjection wells and 77 production wells in this reservoir in Tianjin. All the doublet systems carry on reinjection under free flow condition. All the used geothermal water from the doublets was reinjected into the reservoir after the heating cycle. The amount of reinjection was 2.89×10^6 m^3 in 2007, accounting for about 24% of the total production from the geothermal reservoir. Although there are more geothermal production wells used for space heating, and the adjacent production wells also influence the reservoir pressure around the reinjection wells, it can be observed that the water level close to reinjection wells declines much slower than other parts. The average annual drawdown has been decreasing with the increasing reinjection. On the other hand, there has not been an observable temperature change happen in the surrounding production wells till now. According to the geological condition, a numerical model was set up for the geothermal system in the urban area in Tianjin, using the software package TOUGH2, according to the past 20 years production and reinjection history. The model was used to predict the changes of reservoir pressure of the geothermal system in the future, assuming that ① all the geothermal wells will keep the average production rate in 2002 (80~120 m^3/h in winter, and 5% ~ 10% of the winter production rate in the summer); ② all the 10 reinjection wells are put into use with a reinjection rate of 50~100 m^3/h for each well, and the annual amount of production is $1.371\,6 \times 10^7$ m^3 (deducting the amount of reinjection 1.7×10^6 m^3). It was predicted that the deepest water level in the reservoir will be 193 m below sea level. This means that the sustainability of the geothermal production could not be realized if the present production and reinjection will be maintained into the future. If the present amount of reinjection increases 150%, it was predicted that the deepest water level will be 138 m below see level in 2013. This means that reinjection makes an effective measure to counteract the decline of the reservoir pressure (Wang, 2005).

The reinjected cold water will extract more thermal energy from the rock matrix and improve the heat mining from the geothermal reservoir. But it is not a simple decision to increase the amount of reinjection, because of the possible cooling of the production water. It is proposed that a tracer test be carried out to study the connections between the production and reinjection wells and to predict the cooling effect by the increase of reinjection. In the winter of 1999 and 2001, tracer tests in the dolomite reservoir were carried out and the tracers used including chemical tracer (I, Br^-) and radioactive isotopic tracer (^{125}I, ^{35}S). The distance between the production well and the reinjection well is 850 m. The amount of tracer is 10 kg. The entire tracer was applied instantaneously. The tracer breakthrough time was about 3 d, and the peek time was at about 52 d(See Figure 3). According to the deduction from the tracer tests, there will not be a premature thermal breakthrough.

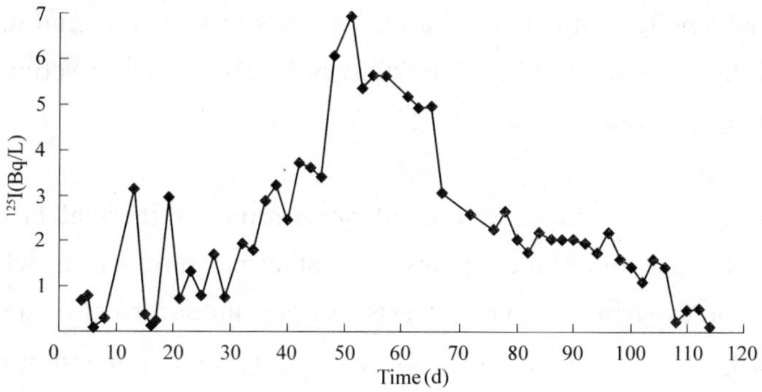

Figure 3 Curve of tracer (^{125}I) recovery of a tracer test

4.3 Sandstone reinjection

By the end of the 1980's, the reinjection tests had been carried out in the Tertiary sandstone reservoir in Tianjin. During the tests, about 30~50 m^3/h waste water were injected into the reservoir. But along with the going on of reinjection, the injectivity decreased quickly. Tests of sandstone reinjection were carried out again in the winter of 2004~2005, and the results were similar to that of the previous tests. The main problems of sandstone reinjection in Tianjin are: ①Injectivity decreases fast with time or injectivity changes too much with time during the reinjection process; ②The clogging of the reinjection wells, attributed to physical, chemical and biological factors.

It is proposed that further tests on sandstone reinjection be carried out, considering the experiences in the oil fields in China and the experiences in the world.

Now some tests and research jobs are progressing steadily, on the aspect of the wellhead facilities, drilling and casing techniques of reinjection wells in sandstone reservoirs.

5 Discussions

5.1 Distance of production and reinjection

The cooling of produced geothermal fluid caused by injection of colder fluid has been reported in a few high-enthalpy geothermal fields. For low-enthalpy geothermal fields, there has not been any such report, even in the cases where the distance between the production well and injection well is rather small. Therefore, it may be concluded that for production-reinjection doublets in low-enthalpy geothermal fields, one does not have to fear about the cooling of the production water, if the distance between production well and injection well is greater than a few hundred meters, and the amount of reinjection in not very huge. In designing the distance of reinjection and production wells of a doublet system, a few factors should be considered, including the type of geothermal reservoir, the geological structure of the geothermal field, the permeability and thickness of the reservoir, the direction of fluid flow, the temperature difference between the reservoir and reinjection water, the flow rate of reinjection etc. But in the cases that a large number of reinjection wells and production wells will be placed among a rather small area, care has to be taken, and proper tests have to be carried out and proper modeling has to be done before any such injection project is started, so as to avoid premature thermal breakthrough.

5.2 Tracer test

Tracer breakthrough can be a very good precaution for thermal breakthrough. Tracer testing is one of the most important aspects of geothermal reinjection, which has become a routine for reinjection experiments. Tracer tests can provide information about the flow paths and the flow velocity of the geothermal fluids between the injection and production wells. For fractured reservoirs, the volume of the aperture can be deduced from the tests. This information can be used to predict the cooling due to reinjection (Axelsson and Stefánsson, 1999). For reinjection projects that are large-scale or a number of production and reinjection wells in a relatively small area, it is strongly proposed that a tracer test be carried out.

5.3 Monitoring

Monitoring is one of the most important elements for geothermal management. For a geothermal field with reinjection, a proper monitoring program is even more important. Besides the monitoring of reservoir pressure, temperature, amount of production, chemical composition of geothermal fluid etc, the water level in the injection wells, temperature of injection water, amount of injection and chemical composition of injection water should also be monitored. The purpose is to find out the changes of the geothermal system caused by reinjection, especially the cooling of the produced geothermal water.

6 Concluding remarks

Reinjection is one of the most important aspects for the sustainable management of geothermal resources. Reinjection tests were started as early as 1974 in Beijing. Scaled reinjection of return geothermal water form heating systems has been applied in Tianjin since 1996 and in Beijing since 2001. The reinjection experiences show that it is significant in controlling the lowering of reservoir pressure, and improves the heat mining of the geothermal field. In the Xiaotangshan area in Beijing, the amount of reinjection has to be about 60% of the geothermal water production. As a result, the water level of the geothermal field has stopped declining. This is a very good example of sustainable use of geothermal resources. It is proposed that reinjection can be expanded in Beijing and Tianjin, as well as other geothermal fields in the world. On the other hand, reinjection is the most complicated technique of reservoir engineering, and premature thermal breakthrough should be avoided, and tracer tests and proper monitoring program should be carried out.

References

1 Axelsson, G., 2003: Essence of geothermal resource management. Edited by Ffidleifsson I.B. and Gunnarsson M.. *Lectures on the sustainable use and operating policy for geothermal resources—Short course prior to the International Conference IGC2003.* United Nations University Geothermal Training Programme, p.249.

2 Axelsson, G. and V. Stefansson, 1999: Reinjection and geothermal reservoir management-associated benefits. *Proceedings of the International Workshop on Direct Use of Geothermal Energy* . Ljubjana, Slovenia, p.21.

3 Liu, J., 1999: *Reinjection and tracer tests in the Laugaland Geothermal field, N-Iceland*, UNU G.T.P., Iceland, pp.141-164.

4 Han Zaisheng, 2002: Artificial Recharge of groundwater in North China Plain, Management of Artificial Recharge for Sustainability, A. A. Balkema Publishers, pp.435-438.

5 Liu, J. and Y. Yan., 2004: Geothermal resources and direct use in the area of Beijing. *Proceedings of the 6th Asian Geothermal Symposium.* KIGAM, Daejeon, Korea. pp.25-31.

6 Liu, J., X. Pan, W. Ran, X. Wang and W. Xu., 2002: A few aspects of geothermal reinjection in Beijing. *Proceedings, 2002 Beijing International Geothermal Symposium,* China Geological Publishing House. pp.373-380.

7 Rybach, L., 2003: Sustainable use of geothermal resources: renewability aspects. Edited by Ffidleifsson I.B. and Gunnarsson M.. *Lectutes on the sustainable use and operating policy for geothermal resources—Short course prior to the International Conference IGC2003.* United Nations University Geothermal Training Programme. p.249.

8 Stefansson, V. and G.Axelsson., 2003: Sustainable utilization of geothermal energy resources. Edited

by Ffidleifsson I.B. and Gunnarsson M.. *Lectutes on the sustainable use and operating policy for geothermal resources—Short course prior to the International Conference IGC2003.* United Nations University Geothermal Training Programme. p.249.

9 Wang K. *et al.*, 2001: *The modeling Research about the deep basement geothermal reservoir in Tianjin(in Chinese).* Tianjin Geothermal Exploration and Development Institute, Report pp.20-30.

10 Wang K., 2005: The Sustainable Development and Utilization of Geothermal Resources in Tianjin, China, *Proceedings World Geothermal Congress 2005.*

Study on Quality Change of Roof Rainwater and Sand Filter Column Effect

Wang Weiping, Sun Xiaobin, Qu Shisong, Xu Yu, Zhang Peipei, Liu Pin
(School of City Development, University of Jinan, Jinan ,250022)

Abstract: Based on pollutants of roof rainwater runoff in Jinan urban area, the paper reveals dynamic regularity of different pollutant varying with rainfall and water quality changing with season. According to pollutants in roof rainwater, a sand filter system was designed to remove pollutants. The test results showed that the system was effective in removing pollutants in roof rainwater, the average removal rate of turbidity reached 87% ,and the average removal rate of suspended solids was above 70% ,the sand column was also effective for purification of chroma, ammonia nitrogen, volatile phenol, nitrite nitrogen, zinc and so on. Filtering test provides basic research for utilization of roof rainwater in Jinan.
Keywords: roof rainwater runoff; pollution characteristics; rainwater quality; test; sand column filtration system

1 The law of rainfall in Jinan

1.1 General situation of precipitation in Jinan City

Jinan is located in east longitude 117°00′30″, north latitude 36°40′6″, the northern foot of Mountain Tai, it is next to Yellow River in the south. It has a continental climate. The weather is dry in spring, hot and rainy in summer, warm in autumn and cold in winter. The genenal precipitation in Jinan is shown in Figure 1. The largest precipitation over years is 1 160 mm (in 1962), the smallest precipitation is 320.7 mm (in 1968), the difference is 839.3 mm, the average annual precipitation in urban area is 671.1 mm (1971~2000), the precipitation in June and July occupies 57% of annual rainfall, rainfall from June to September accounts for about 85% of the year, which represents flood season.

1.2 Daily maximum precipitation and maximum rainfall per hour in Jinan

Figure 2 shows daily maximum precipitation and maximum rainfall per hour in Jinan. Since 1961, there have been 18 years that the daily maximum precipitation is more than 100 mm in urban area, and there have been 6 years that daily precipitation is more than 150 mm, the maximum amount of daily precipitation is 298.4 mm (in 1962). There are 15 years that maximum rainfall per hour is more than 50 mm,and there are 2 years that maximum rainfall per hour is more than 100 mm.The maximum rainfall per hour is 151.0 mm (in 2007), Figure 2 shows the daily maximum precipitation and maximum rainfall per hour in Jinan urban area.

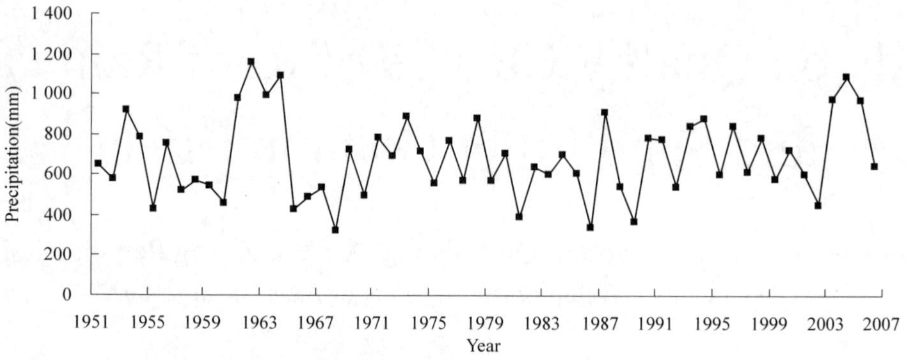

Figure1　Annual precipitation curve in Jinan urban area

(Source: Jinan Meteorological Bureau)

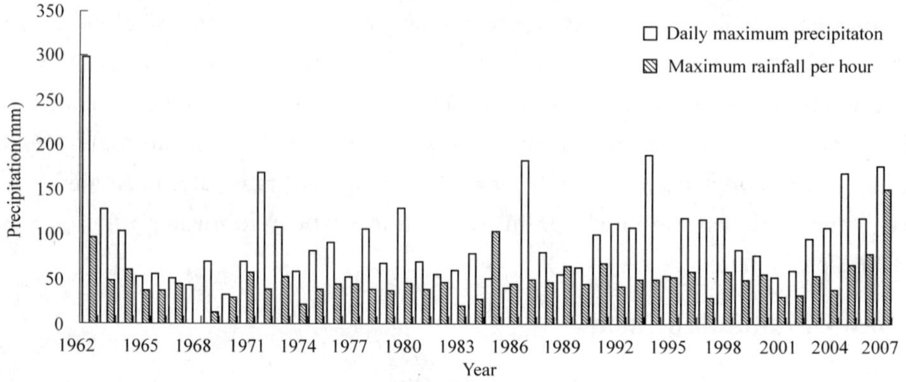

Figure 2　Daily maximum precipitation and maximum rainfall per hour in Jinan urban area

(Source: Jinan Meteorological Bureau)

1.3　Summer precipitation in Jinan

Figure 3 shows the amount of Summer precipitation in Jinan, which reflects rainfall fluctuating among years.

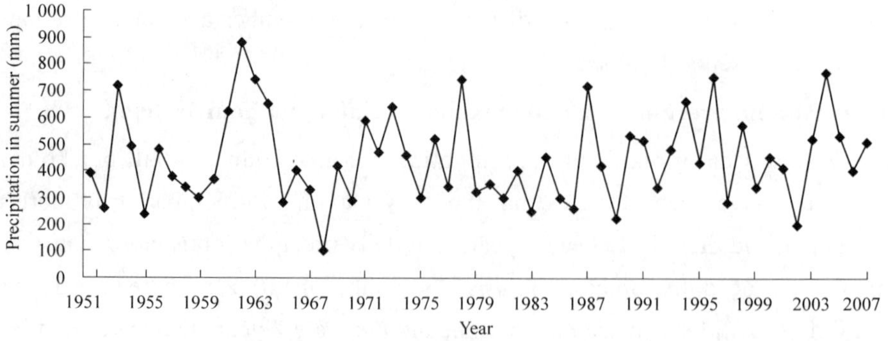

Figure 3　Summer precipitation changes curve in Jinan urban area

(Source: Jinan Meteorological Bureau)

2 Quality change of roof rainwater runoff

2.1 Water quality distribution of rainfall with temporal and spatial

Rainwater quality is uneven distribution with temporal and spatial. There are 3 rain water quality monitoring stations in environmental monitoring stations of Jinan, which are Changqing District, Tianqiao District and Lixia District, and monitoring indicators conclude pH, conductivity, sodium ion, potassium ion, calcium ion, magnesium ion, fluoride compounds, nitrate ion, chloride ion, sulfate ion and ammonium ion. According to analysis of 13 water samples in 2008, the major pollutants such as nitrate ion, sulfate ion, ammonium ion and conductivity had character as follows: the maximum concentration of ammonia ion was 6.52 mg/L, the minimum was 1.18 mg/L, it reveals that the ammonia ion concentration of all the samples was beyond the standard for groundwater quality or surfacewater environmental quality standards for Class Ⅲ, and the other indicators did not exceed. In Tianqiao District, the nitrate concentration in summer was 2.5 times than that in spring, while in Changqing District it was 1.6 times.

2.2 The relationship between pollutants concentration of rainwater on the roof and rainfall duration or intensity of rainfall

The sampling site is No.17 Building for teachers in West Campus of University of Jinan, the material of roof is platform which is waterproof, and bottom material is covered by polyurethane, above the roof it covers SBS (modified asphalt) and sand, on top of roof it covers cement mortar; then tile is covered on slope roof. And we collect water samples at interval of 10 min, 15 min, or 30 min.

Roof rainwater runoff is different with natural precipitation; the pollutants of roof rainwater are affected not only by air quality, but also the number of pollutants on roof, roofing materials, roof structure, and the wind direction and so on. According to analysis of monitoring results, turbidity is associated with varieties of pollutants in roof rainwater, on behalf of the analysis we take turbidity as a eigenvalues.

2.2.1 The relationship between pollutants of roof rainwater and rainfall duration

Researching results reveal pollutants of roof rainwater runoff varies with time and rainfall intensity plays an important role in determining to discard the volume of initial rainfall and recharging karst aquifer. Because pollutants concentration of roof rainwater runoff are not only affected by rainfall duration, but also by rainfall intensity. During a rainfall the more serious rainfall intensity is, the higher concentration of pollutants is, and most of that happens in early stage of rainfall; and in latter process of precipitation as rainfall intensity increases, but the concentration of pollutants will reduce.

It is hard to find that disciplinarian when rainfall precipitation is little, and the longer rainfall duration lasts, the phenomenon will easily happen in early stage of precipitation. When the rain gets started, at first pollutants concentration would increase, as rainfall intensity continues increase, the pollutants concentration would increase too, and after a while it reduced. This phenomenon is obviously reflected on March 12 with the rainfall precipitation is 2.6 mm, and the rainfall duration is 3 h and 20 min, if the sampling time makes a great difference, the difference cannot be seen easily.

2.2.2　The relationship between pollutants of roof rainwater and rainfall intensity

The relationship between concentration of suspended solids and rainfall intensity on July 1 will be discussed in this part. At initial rainfall, turbidity of 3 mm roof rainwater was high, until 8 mm the rainwater quality began to get good, but there were still coarse-grained sediments (such as wall skin or particle). Types of suspended solids were different due to variation of rainfall intensity. At the beginning of rainfall, rainfall intensity was certain, the type of suspended solids was also certain, when the rainfall intensity increased the concentration of suspended solids would increase until most pollutants were washed away and then rainwater would get clean, after 8 mm precipitation the amount of fine particles was a little, mainly left coarse-grained.

3　Quality changes of roof rainwater in Jinan during a year

Pollutants of roof rainwater runoff have great changes in each rainfall. From spring to summer turbidity of roof rainfall tends to decrease, and in summer the value is a little, even can reach good condition. For example, on July 1 turbidity of the early roof rainwater runoff was 4 NTU.

The first rainfall in spring happened on March 12, while pollutants of the roof rainwater runoff were the most serious, so it was not a wise choice to recharge using this water. Testing results: COD value was 23.0 mg/L, NH_3–N value was 29.5 mg/L, and turbidity value was 3 500 NTU. Compared with pollutants of rainfall in spring, pollutants concentration of roof rainwater in summer was less, COD value was 13.6 mg/L, NH_3–N value was 0.93 mg/L, and turbidity value was 160 NTU.

4　Filter equipment, process and methods

4.1　Test device

Filter column device is made of polymethyl methacrylate. Area of cross section is 0.013 3 m^2, the height of filter supporting bed is 10 cm, with thickness of quartz sand (size of 1~2 mm, 0.25~0.5 mm ,0.08~0.25 mm) is 45 cm, 30 cm, 45 cm separately. The average permeability coefficient of sand column is 37.21 m/d. Figure 4 Shows Test Device.

4.2　Test process

Test water samples was collected from precipitation in spring and summer ,and in first experiment the sand column was filled with diameter of 1~2 mm, 0.25~0.5 mm, 0.08~0.25 mm with thickness of 45 cm, 30 cm, 45 cm separately. In initial phase of water supply, rainwater was transferred from tank through pipe and rotameter into the sand column and water head should be maintenanced at 50 cm, while water head achieved stability, accommodated the rotameter at 29 L/h. We got water sample every 30 min from lowest tap of every sand layer. Every experiment we could take 10 bottles of water (include source water). In second experiment a sand layer will be added on top of sand column with diameter of 2~3 mm and thickness of 10 cm.It was defined as managed layer, which may have great significant for the removal of pollutants.

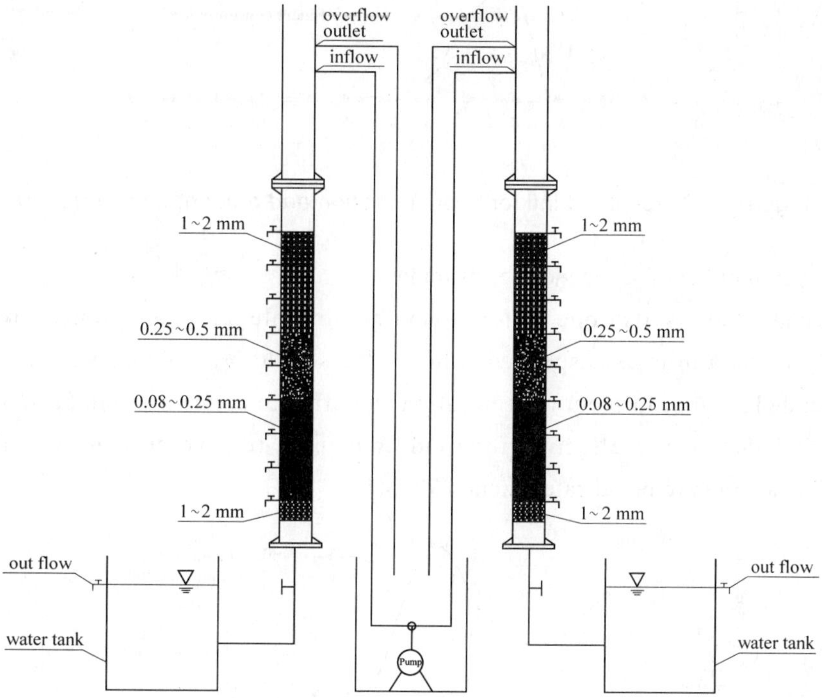

Figure 4 Test device

5 Test results

5.1 The overall effect of pollutant removal

According to the detection of roof rainwater, the precipitation in spring should not be available. The compositions of pollutants which exceeded groundwater quality standard concluded ammonia nitrogen, nitrite nitrogen, suspended solid, chroma, turbidity, volatile phenol, lead (groundwater quality standard for Class Ⅲ).

5.1.1 The overall effect of turbidity removal

Turbidity is one major reason causing plugging which is a serious problem for recharge well and aquifer, national drinking water standard (old) ordain turbidity value should be less than 3 NTU. Figure 5 shows turbidity of influent concentration and effluent concentration. From the figure it is clear that sand filter has good impact on turbidity removal. The average removal rate reached 87.84 %. The value of effluent concentration met the national drinking water standards (old),even for pollutant with turbidity more than 60 NTU the removal efficiency was close to 100%, for pollutant with turbidity less than 60 NTU its removal efficiency was 83.69%.

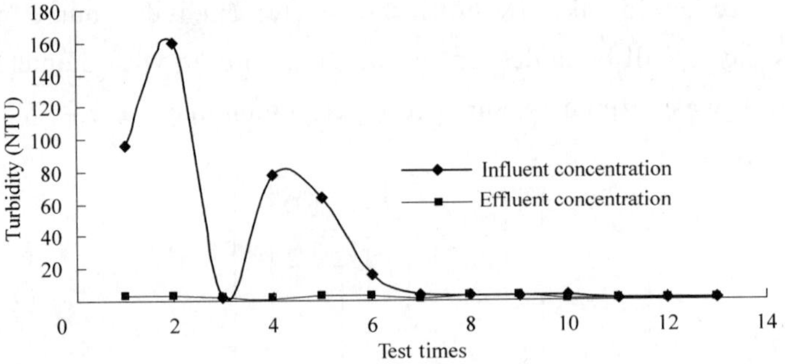

Figure 5 Turbidity of influent concentration and effluent concentration

5.1.2 The overall effect of suspended solids removal

Suspended solids is also one major reason causing plugging, and water environmental standard of Australia ordains suspended solids value should be less than 20 mg/ L, Figure 6 shows suspended solid of influent concentration and effluent concentration. From the figure it is easy to find that it was effective for sand column to remove suspended solids in roof rainwater. The average removal rate reached 77.24%.

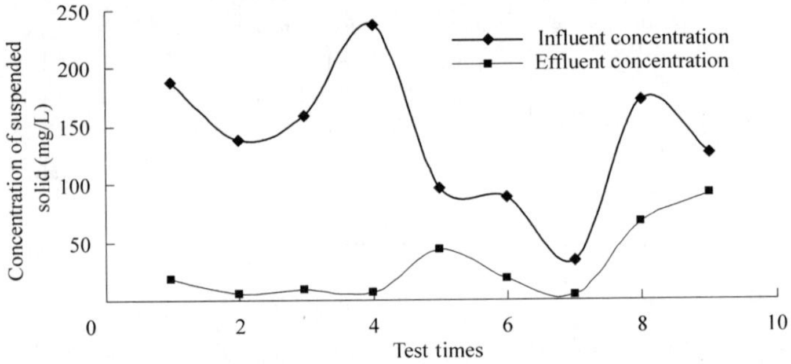

Figure 6 Suspended solid of influent concentration and effluent concentration

5.1.3 The overall effect of ammonia nitrogen removal

National drinking water standards ammonia nitrogen should be less than 0.2 mg/L, Figure 7 shows ammonia nitrogen concentration of influent and effluent, the average of removal rate was 57.63%, and if ammonia nitrogen concentration in rainwater was high, then removal rate would be reduced, for example ,when its concentration was more than 1 mg/L, the removal rate of ammonia nitrogen was 25.6%, which could not reach the value of drinking water standard. While the concentration was less than 1 mg/L, the average removal rate reached 71.86%, after filtering process the ammonia nitrogen value could achieve the drinking water standard. So the higher concentration of ammonia nitrogen was, the lower removal efficiency of sand column was.

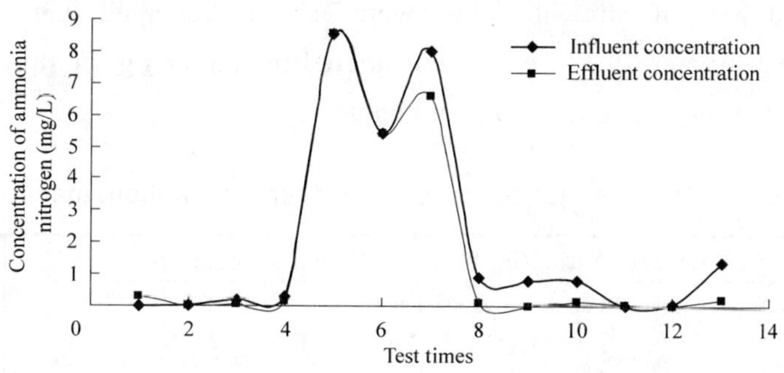

Figure 7 Ammonia nitrogen of influent concentration and effluent concentration

5.2 The removal efficiency for managed layer

On top of the sand column plus diameter of 2~3 mm with thickness of 10 cm, aims to remove large particle suspended solids and other pollutants in initial rainwater, and at the same time, in practice the layer on the top could be managed by replacement.

5.2.1 Effect of turbidity removal for managed layer

From Table 1 and Table 2, it is easy to find that it was obviously for managed sand layer to remove turbidity, because sand column was already effective to remove turbidity without managed layer.

Table 1 The effect of turbidity removal without managed layer

Test times	Influent concentration(NTU)	Effluent concentration(NTU)	Removal rate(%)
1	96	3	96.88
2	160	3	98.13
3	64	3	95.31
4	16	3	81.25

Table 2 The effect of turbidity removal with managed layer

Test times	Influent concentration(NTU)	Effluent concentration(NTU)	Removal rate(%)
1	77.4	1.2	98.45
2	3.6	1.1	69.44
3	3	3	—
4	3	3	—

5.2.2 The removal effect of suspended solid with managed layer

Table 3 and Table 4 show the differences of removal efficiency between no managed layer and plus managed layer, we could find the average of removal efficiency of suspended solids was 80.44% without managed layer, and while there was a managed layer on top of sand column, the average of removal efficiency raised to 85.22%.So we could draw a conclusion that adding management layer was helpful to raise the removal efficiency of suspended solid in roof rainwater;. however, Table 4 revealed that removal rate would decrease as the test went on , so it was considered for replacing the upper managed layer in order to slow down removal burden of sand column.

Table 3 The effect of suspended solid removal without managed layer

Test times	Influent concentration(mg/L)	Effluent concentration(mg/L)	Removal rate (%)
1	187	18	90.37
2	138	5	96.38
3	96	43	55.21
4	89	18	79.78

Table 4 The effect of suspended solid removal with managed layer

Test times	Influent concentration(mg/L)	Effluent concentration(mg/L)	Removal rate (%)
1	159	9.6	93.96
2	236	6.2	97.37
3	33	3.8	88.48
4	172	67	61.05

5.2.3 Effect of ammonia nitrogen removal with managed layer

From Table 5 and Table 6, we could draw a conclusion: it is obviously that adding managed sand layer could raise removal efficiency of ammonia nitrogen. The average of removal efficiency would reach 58.85% .

Table 5 The effect of ammonia nitrogen removal without managed layer

Test times	Influent concentration(mg/L)	Effluent concentration(mg/L)	Removal rate (%)
1	0.01	0.01	—
2	0.01	0.01	—
3	8.54	8.54	—
4	5.44	5.44	—

Table 6 The effect of ammonia nitrogen removal with managed layer

Test times	Influent concentration(mg/L)	Effluent concentration(mg/L)	Removal rate (%)
1	0.2	0.066	67.00
2	0.31	0.15	51.61
3	8.02	6.57	18.08
4	0.78	0.01	98.72

6 Conclusions

First part of this article analyzes the characteristics of rainfall in Jinan, we draw a conclusion that pollutants concentration of roof rainfall in spring is highest, so we should abandon all of this rainwater, and we also explore dynamic regularity of different pollutant varying with rainfall and water quality changing with season, all of these rescarch will provide a basis theory for utilization of roof rainwater in Jinan urban area. And on second part of paper based on an experiment, mainly study on removal efficiency of sand-filter system which was composed with different size of sand and the purification, here are the conclusions:

(1) Sand-filter system can be used to treat pollutants of roof rainwater runoff; it is effective to remove turbidity, suspended solids, chroma, volatile phenol, lead ion, zinc ion, arsenic and mercury in rainwater. And average removal rate of suspended solids was above 70%, the average removal rate of turbidity reached 87.84%, and the average removal rate of chroma was more than 50%, and after treatment the value of turbidity, chroma, lead ion, zinc ion, arsenic and mercury in rainwater could achieve the value of drinking water standard. So sand-filter system is effective to remove some pollutants in roof rainwater.

(2) It is clear that the effect of pollutants removal mainly happens in upper sand column, the most serious plugging happens on top of sand column. And it leads to the decreasement of permeability coefficient of the sand column, the average penetration coefficient of sand filer system decreases from 37.21 m/d to 26.63 m/d.

(3) Adding managed layer on top of sand column is more effective to remove suspended solids, chroma and ammonia nitrogen, and removal efficiency of suspended solids can be stabilized at 80%, with running time increases the removal efficiency will decline, which requires top managed layer must be replaced at intervals.

Groundwater Recharge with Municipal Effluent in China

Zhao Xuan, Cheng Xuzhou, Zhang Meng
(Division of Environmental Technology, Institute of Nuclear and New Energy Technology, Tsinghua University, Beijing, P. R. China, 100084)

Abstract: In China the first groundwater recharge demonstration plant with reclaimed municipal wastewater was setup and devoted to operation in 2002 with a capacity of 500 m^3/d in Beijing Gaobeidian WWTP. This paper summarized the results and experiences of one year period operation. The aquifer treatment contributed 40% removal of the bulk organics in terms of DOC, COD, and UV254 absorbance. The recovered water quality meets requirements of non-potable purposes.
Keywords: municipal effluent; artificial groundwater recharge; reclaim; reuse

1 Introduction

China is a country with water shortage and uneven water resource distribution. More than 400 cities in China face water shortage problem. Beijing, with 14 million inhabitants, is a typical Chinese city that is very short of water resources. Per inhabitant possess of only 300 m^3 of water available per year as compared to 2 200 m^3 in China or even 9 000 m^3 worldwide. Since two thirds of Beijing's water supply is taken from groundwater, the groundwater resource is, inevasible, over exploited and the groundwater levels dropped from 25 m in 1998 to 19 m in 2002. The annual drop amounts to 1.6 m.

Municipal wastewater reclamation for artificial groundwater recharge provides one of the possibilities of water supply supplement, together with other options such as south-north water transfer project and sea water desalination. In 2004, the secondary treatment capacity of the wastewater system in Beijing reached 1.88×10^6 m^3/d, with the treatment ratio of 86.4%. In 2007 the total municipal wastewater will be treated secondarily with the capacity of 2.68×10^6 m^3/d. The possibility, therefore, does exist in Beijing to apply artificial recharge technology to acquire high quality reclaimed water for severe groundwater level drop and high treatment efficiency of municipal wastewater.

Artificial groundwater recharge, with municipal effluent, exhibits the favorable advantages such as additional natural treatment, relief of groundwater level dropping, storage capacity to buffer seasonal variations of supply and demand, etc. As experienced with the full scale projects in the U.S., Europe, Israel, Australia and so on (T. Asano *et al.*, 2004).

Surface spreading and direct aquifer injection are the two types of techniques most commonly used in applying artificial groundwater recharge with reclaimed municipal wastewater. Pretreatment requirements for groundwater recharge vary greatly depending upon the purpose of specific groundwater recharge project, the source of reclaimed wastewater, the type of recharge method and also public acceptance. For surface spreading with soil aquifer barrier, the recommended pretreatment for groundwater recharge usually are secondary and tertiary treatment (eg flocculation, sand filtration and disinfection) as applied in Israel, the U.S., Spain, Italy, Belgium, and Australia(T. Asano *et al.*, 2004; Kanarek *et al.*, 1996; Drewes, 1996; Dillon, 2000). In the cases of direct aquifer injection or aquifer recharge for domestic water supply(eg in California and Belgium) intensive treatment such as the title 22 technology and double membrane systems was implemented (T. Wintgens *et al.*, 2005).

Artificial groundwater recharge can contribute in a sustainable way to the stabilization of water supply situation. However, in China, there is still no such practice in the aim of water recycling and reuse within the framework of integrated water resource management. Since 1996 Tshinghua University and Beijing Drainage Corp. has been working together with the Technical University of Berlin on the problem of indirect potable reuse for groundwater recharge. In 2001, the first pilot plant in China related with reclaimed municipal wastewater for groundwater recharge was constructed in Gaobeidian WWTP.

The aim of the case study was to investigate the fate of the bulk parameters in the stages of pretreatment and the soil aquifer treatment and optimize the operation strategy, thus to provide a sound basis for the implementation of full scale groundwater recharge projects.

2 A case study in Beijing

2.1 General description

The artificial groundwater recharge pilot plant is located within the area of Gaobeidian Wastewater Treatment Plant (WWTP). Gaobeidian WWTP is the largest wastewater treatment plant in China, with a capacity of 1 000 000 m^3/d, that occupies about 40% of the total sewage discharge in Beijing. The plant undertakes the task of wastewater treatment for a total catchment area of 9 661 hm^2, ranging from the city centre to the east suburb of Beijing, with the inhabitants of more than 2.4 million. The treatment process of the plant is conventional activated sludge secondary treatment.

Before construction of this demonstration plant, a field investigation was made including historic data survey and geological drilling test, for analysis of physical, chemical and hydraulic properties of soil samples at different depths. Table 1 summarizes the results from the investigation survey.

Table 1　Drilling test results at the Gaobeidian WWTP aquifer recharge site

depth (m)	Lithology description	aquifers
0 ~ 9.6	surface layer is filling soil, down layer is powdery clay, viscid powdery soil with a thin layer powdery sand	aquifer Ⅰ
9.6 ~ 14.0	silver sand, silver medium sand	
14.0 ~ 17.5	round gravel	
17.5 ~ 24.8	powdery clay, sandy powdery soil	
24.8 ~ 28.0	silver sand	aquifer Ⅱ
28.0 ~ 33.2	medium, coarse sand in upper layer and cobble stone in down layer	
33.2 ~ 35.0	clay	
35.0 ~ 39.4	silver sand, medium silver sand	
39.4 ~ 45.0	powdery clay	
45.0 ~ 50.5	mainly round gravel with silver sand　in top and bottom	aquifer Ⅲ
50.5 ~ 56.0	clay and powdery soil alternant	

In view of the poor permeability of the unsaturated zone, the pilot plant was designed and constructed as shown in Figure 1 and Figure 2.

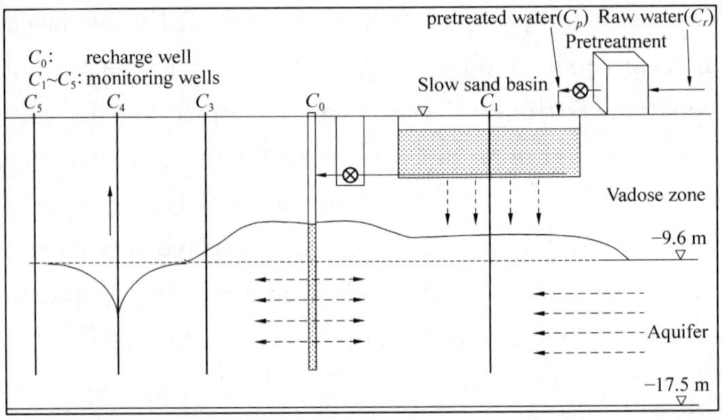

Figure 1　Sampling points for the Gaobeidian WWTP case study site

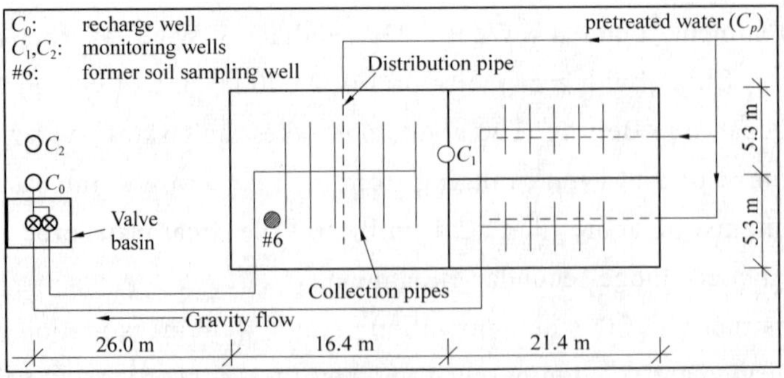

Figure 2　Sampling points for the case study site in relation to the sand infiltration basin

The secondary effluent from Gaobeidian WWTP is firstly dosed with 30 mg/L PAC, and then goes through sedimentation and fast sand filtration for SS removal. The effluent is then filtrated with GAC before being diverted to the slow sand infiltration basin for recharge. The sand infiltration basin has a 1 m deep sand layer and water distribution–collection pipes system. The water enters the recharge site and passes a sand layer. Then drainage pipes divert the water to the infiltration wells with a completely filtered section between 9.6 m and 17 m. The drainage pipes are covered with a felt layer to prevent clogging sand particles. The basin is filled with treated water up to a certain level. Then the water level decreases until it reaches the height of the effluent pipe.

2.2 Results and discussion

Since 2002, Beijing Drainage Corp. has been running this plant on the site of the sewage plant. The infiltration basin was operated with 7 ~ 10 d wet/dry cycle in order to control eutrophication and the formation of clogging layers. The recharged capacity amounts to 500 m^3/d. The reclaimed water quality had been monitored 2 times per month for one year. In addition, the original groundwater quality was monitored for three months before recharge. Figure 3 ~ Figure 5 summarized the reclaimed water and groundwater qualities.

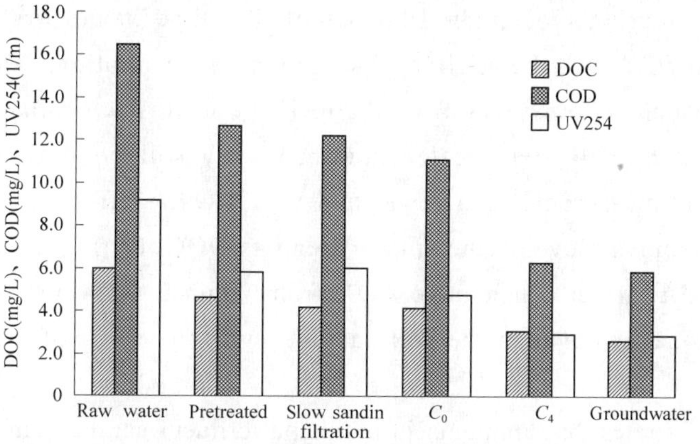

Figure 3 organics removal efficiency

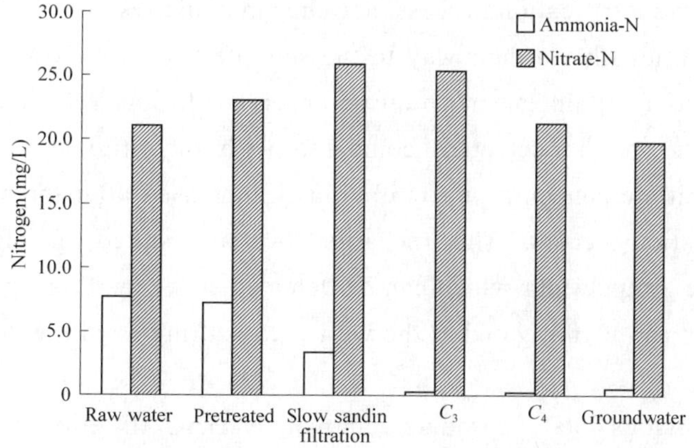

Figure 4 Ammonia–N and nitrate–N removal

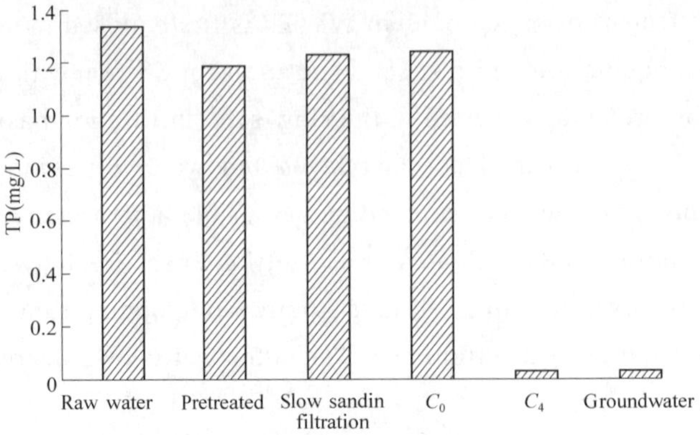

Figure 5　TP removal

The pretreatment process, flocculation and GAC filter, can reduce the DOC value of the reclaimed water from 5.99 mg/L to 4.61 mg/L. The slow sand basin, as the protection barrier of recharge well from clogging, did not exhibit the performance of organic removal. By the aquifer passage, which serve as the aerobic, anoxic, and anaerobic bio-reactor in sequence, about 2 mg/L DOC (approx.40% of content in effluent after pretreatment) of the reclaimed water was removed during more than 10 d percolation time in the first aquifer, and thus values of about 3 mg/L DOC, the aquifer's background concentration, can be reached at the monitoring point which is 26 m away from the recharge well. The similar DOC removal had already been detected in the lab scale experiments by soil columns (Ernst, 2000) and confirmed with field measurements (Drewes and Fox, 1999; Steffen Grunheid *et al.*, 2005). COD and UV254 removal develop the similar trend as DOC. During the operation period it was detected that GAC filter can remove AOX from 70 μg/L to 54 μg/L. However, there is not obvious additional removal in the soil passage, with the value of 50 μg/L AOX in the recovered water.

Figure 4 demonstrates that ammonia in municipal effluent can be completely transformed into nitrate by the artificial recharge process shown in Figure 1. The slow sand basin performed an effective nitrification process, and the ammonia residues in the reclaimed water can be removed additionally on their way to the saturated zone. However, denitrification in the saturated zone to a certain extent could be observed. In the well of C_3 and C_4 elevated nitrate concentration can be detected, compared with the original concentration in the groundwater. The nitrate concentration is obviously increased after the slow sand basin, in which the nitrification occurs. The recharged water resulted in the elevated nitrate concentration in the groundwater which can be detected at the monitoring well C_3. However, this negative impact can be mitigated at the well C_4, resulting from the denitrification in the saturated zone.

There is no obvious total phosphate removal by GAC filtration and sand infiltration steps. However, the soil aquifer itself performed effective TP adsorption, since the elevated concentration of TP can not be detected in the recovered water from well C_4. Despite the relatively short soil passage the total concentration of phosphorus is reduced to less than 100 μg/L (removal rate>90%) by adsorption in the soil body.

The recovered water from artificial groundwater recharge with reclaimed wastewater could be used for non-potable reuse options, such as agriculture irrigation, urban irrigation (golf courses, sports fields), power plant cooling, industrial processing, construction, toilet flushing, wetland construction, and so on. Potable use is often not recommended because of high residual organic carbon contents, consisting of a broad spectrum of mostly synthetic organic chemicals. Regulations vary throughout the world. In this aspect careful risk assessment should be carried out.

3 Conclusions

The demonstration artificial groundwater project exhibited promising purification effects of reclaimed municipal effluent in terms of bulk parameters including soluble organics, nitrogen and phosphorus. The reclaimed water meets the water quality requirements of various non-potable applications. In a word, the study brought us an optimistic scenario of the artificial groundwater recharge application in China in the near future.

References

1 Cao Honglin. The Macro-analysis on Target of Sewage Treatment Rate and Related Non-target for the Olympic Games (in Chinese). *Municipal Engineering and Technology*, Vol. 24, No. 4, July, 2006: 217-219.

2 M. Ernst, A. Sachse, C. E. W. Steinberg, *et al.* (2000). Characterization of the nanofiltration permeates of a tertiary effluent, *Wat. Res.* Vol. 34, No. 11, pp. 2879-2886.

3 Kanarek, A. , Michail M. (1996). Groundwater recharge with municipal effluent: Dan Region reclamation project. *Israel, Water Science Technology*, 34 (11), pp. 227-233.

4 Shelef G., Azov Y., (1996). The coming era of intensive wastewater reuse in the Mediterranean region. *Water Science Technology*, 33(10/11), pp. 115-125.

5 T. Wintgens, T. Melin, A. Schaefer, *et al.* (2005). The role of membrane processes in municipal wastewater reclamation and reuse, *Desalination*, 178(2005), pp. 1-11.

6 Takashi Asano, Joseph A. Cotruvo (2004). Groundwater recharge with reclaimed municipal wastewater: health and regulatory considerations, *Water Research*, 38 (2004), pp. 1941-1951.

7 Jörg E. Drewes, Peter Fox (1999). Behavior and Characterization of Residual Organic Compounds in Wastewater Used for Indirect Potable Reuse, *Water Science Technology*, 40(4-5), pp. 391-398.

8 Drewes J., Jekel M.(1998). Behavior of DOC and AOX using advanced treated wastewater for groundwater recharge. *Water Research*, 32(10), pp. 3125-3133.

9 Drewes J., Fox P.(1999). Fate of natural organic matter (NOM) during groundwater recharge using reclaimed water. *Water Science Technology*, 40 (9), pp. 241-248.

10 Ernst M., Jekel M.(1999). Advanced treatment combination for groundwater recharge of municipal wastewater by nanofiltration and ozonation. *Water Science Technology*, 40(4/5), pp. 277-284.

11 Dillon P. (2000). Water reuse in Australia: current status, projections and research. *Proc. Water Recycling Australia 2000. Adelaide*, 19-20 Oct 2000 (ed. P.J. Dillon), pp. 99-104.

12 Steffen Grunheid, Gary Amy, Martin Jekel. Removal of bulk dissolved organic carbon (DOC) and trace organic compounds by bank filtration and artificial recharge, *Water Research*, 39(2005), pp. 3219-3228.

山东省地下水开采存在问题与对策

马振民，张亮

(济南大学，济南，250022)

摘 要：山东省水资源总量为 363.7 亿 m^3，地下水资源总量为 165.5 亿 m^3，地下水可开采资源量为 159.8 亿 m^3/a，地下水开采量为 131.0 亿 m^3/a。地下水开采引起地面沉降、岩溶塌陷、沉降裂缝等地质环境问题；水资源减少、水质恶化、海水(咸水)入侵、地下水超采漏斗等水环境问题；以及植被退化、土地盐渍化等生态环境问题。建设地下水库、实施地下调蓄、增源增采、涵养地下水源、浅深开采适度、咸淡利用结合等是解决地下水开采存在问题的基本对策。

关键词：地下水；开采；问题与对策；山东省

1 山东省地下水资源概况

1.1 山东省水资源量

山东省位于中国东部，东邻渤海和黄海，中部为低山丘陵，西部和北部为平原，面积 15.67 万 km^2，全省多年平均水资源总量为 363.7 亿 m^3，占全国水资源总量 27 115 亿 m^3 的 1.3%。

山东省地下水资源总量 165.5 亿 m^3，占水资源总量的 45.5%。

山东省地表水资源总量 198.2 亿 m^3，占水资源总量的 54.5%。

山东省每平方千米土地拥有水资源量约为 23.2 万 m^3，仅占全国每平方千米土地拥有水资源量的 2/3。

全省人均占有水资源量 386.5 m^3/a，只有全国人均占有水资源量的 1/6。山东省人口呈增加趋势，随着人口的增加，人均水资源将会进一步减少。

1.2 山东省水资源可利用量

山东省多年平均水资源可利用量为 265.5 亿 m^3/a，水资源可利用率为 73%。人均占有水资源可利用量仅 282.4 m^3/a。

全省多年平均地下水补给资源量为 198.99 亿 m^3/a，地下水可开采资源量为 159.8 亿 m^3/a，占水资源可利用总量的 60%。

全省地表水可利用资源量为 105.7 亿 m^3，占水资源可利用总量的 40%。

2 山东省地下水开采现状

山东省分为 3 个水文地质区，即鲁西北平原以松散岩类为主的水文地质区，鲁中南中低山丘陵以碳酸盐岩类为主的水文地质区，鲁东低山丘陵以松散岩、碎屑岩、变质岩类为主的水文地质区。全省近年平均地下水开采量为 131.0 亿 m^3/a，占地下水可开采资源量的 82.0%。其中，22.1%为工业用水、62.9%为农业用水，生活用水仅占 15.0%。

3 山东省地下水开采存在问题

地下水开发利用不恰当引起的环境问题主要有三类：地质环境问题，水环境问题，生态环境问题。

3.1 地下水开采引起的地质环境问题

地下水资源开发引起的典型地质环境问题包括：地面沉降，岩溶塌陷，沉降裂缝，沼泽化等。

3.1.1 地面沉降

地面沉降是目前世界上许多取用地下水的平原井灌区共同面临的严重问题。超量开采地下水，导致地下水位大幅度下降，促使上部易压缩黏性土层中的孔隙水排出，进而引起土层的固结压缩，这是地面沉降最常见的原因。山东省地面沉降主要分布在鲁北、鲁西南平原的德州、济宁、东营、滨州、菏泽等深层孔隙地下水集中开采区，并受开采水位的控制。

3.1.1.1 山东地面沉降实例：德州市

德州市城区因超量开采孔隙水，20 世纪 70 年代后期已发生地面沉降，随着城区深层地下水长期过量开采，地面沉降的范围已不断扩大，地面以每年 34.1~66.1 mm 的速率下沉，沉降量逐渐增加。据 2007 年年底资料统计，沉降量大于 250 mm 的面积已达 4 438 km^2，沉降区最低点最大累计沉降量为 1 081 mm。

3.1.1.2 山东地面沉降实例：济宁市

济宁市城区位于汶泗河冲积平原的前缘地带，孔隙水开采多年来长期处于超采状态，1988 年以来发生了地面沉降。城区地面沉降值大于 60 mm 的范围已超过 200 km^2。

3.1.2 岩溶塌陷

碳酸盐岩分布区的地下发育有大量洞穴和管道等岩溶形态。其直接被厚度较小的第四纪松散堆积物覆盖的地段由于强烈抽取岩溶水发生覆盖层陷落、开裂、下沉，形成各种规模和形态的地表坑、槽，这种现象称为岩溶塌陷。

山东省岩溶塌陷主要发生在泰安、枣庄、莱芜、临沂等市岩溶水水源地及其周边地区。岩溶塌陷活动频繁，破坏强度大。

山东岩溶塌陷实例：泰安 30 km^2，枣庄 25 km^2。

3.1.3 沉陷裂缝

因过量开采地下水引起地面沉降过程中的岩土体开裂而成。

地面沉陷裂缝实例：广饶县稻庄，长 1 200 m，宽 0.2~2.0 m；济南市长清区南王庄，长 280 m，宽 0.2~0.4 m；泰安市旧县，长 360 m，宽 0.03~0.06 m；淄博市大徐庄，长 200 m，宽 0.2~0.4 m；枣庄市峄城，长 1 450 m，宽 0.02~0.1 m；济南市历城区张马屯，长 20 m，宽 0.05 m。

3.2 地下水开发引起的水环境问题

3.2.1 水资源减少

水资源不合理开发造成的水资源短缺主要有 3 种情况：

(1)地表水、地下水补给资源的减少，例如济南泉域等。

(2)过量开发造成局域缺水，例如淄河流域等。

(3)水质污染使可用淡水减少，例如淄博大武水源地、莱芜清泥沟水源地等。

大规模开发利用水资源，改变了局域水文循环的状况，在某些地区衍生的水循环，会使补给资源量减少。山东省随着土地利用程度的增强，水资源开发力度增大，使用量增多，降水量发生明显变化，20 世纪 80 年代以来，山东省地下水补给总量减少了近 20%。

3.2.2 水质恶化

强烈抽取地下水会改变天然水的水动力条件，形成新的水文地球化学环境，若水中某些化学、微生物组分含量不断增多，就会产生水质恶化现象。

全省平原区面积 73 720 km^2，地下水以Ⅴ类标准为主，其所占面积占全省平原区总面积的 65.6%。

全省山丘区面积 79 476 km^2，地下水符合Ⅲ、Ⅳ、Ⅴ类标准的面积分别占全省山丘区总面积的 30.3%、37.0%、29.6%，符合Ⅱ类标准的面积仅占全省山丘区总面积的 3.1%。

地下水水质恶化的特点是：

(1)地下水普遍遭到 NO_3–N 污染，并成为地下水环境面临的主要问题之一。

(2)地下水中宏量组分含量普遍升高，其中 Cl^-、SO_4^{2-}、Ca^{2+}、Na^+、Mg^{2+}等增幅最快，矿化度、硬度快速上升，地下水水化学类型已发生显著变化。

(3)地下水已遭受 Cr、酚、CN、As、Hg 等人为环境物质的污染，地下水环境受到人类经济技术活动的干扰和影响。

3.2.3 海水(咸水)入侵

滨海地区人为超量开采地下水，降低了陆地地下水位，引起海水(咸水)沿含水层向陆地方向潜移。海水入侵导致陆地淡水含水层咸化、机井报废、土壤盐碱化等。山东省渤海、黄海沿岸不少地区都有海水入侵问题，主要是由于地下水的强烈开采所造成的。

山东省沿海东营、潍坊、烟台、威海、日照、青岛等市共有 19 个县(市、区)发生海水(咸水)入侵，面积已超过 3 100 km^2。其中，最严重的是莱州湾南侧，海水入侵最远距离达 45 km，面积 2 500 余 km^2，目前仍以 70~80 km^2/a 的速度扩展。

海水(咸水)入侵已给当地人民群众的正常生活带来严重影响，给工业、农业造成了巨大的经济损失。

3.2.4 地下水超采漏斗

开采地下水使开采区地下水位下降，改变了地下水的天然流场，形成地下水位降落漏斗。

地下水的持续严重超采，使地下水漏斗区不断扩大。因过量开采深层地下水，在德州、滨州、聊城、济宁、菏泽等地区形成地下水超采漏斗，山东省平原地区浅层地下水漏斗区面积已逾 2.0 万 km^2。

山东地下水超采漏斗实例：德州市德城区超采漏斗。

德州市德城区超采漏斗形成于 20 世纪 70 年代，并且随着地下水开采量逐年增大，地下水位不断下降，漏斗面积也逐年扩大，1990 年漏斗面积 1 707.5 km^2，2000 年漏斗面积 2 552.5 km^2，漏斗面积扩展速率 84.5 km^2/a，仅 2006 年德州城区深层水超采漏斗较上年扩大 510 km^2。目前德城区漏斗中心水位埋深已逾 120 m，地下水超采漏斗面积超过 4 000 km^2。

3.3 地下水开发引起的生态环境问题

3.3.1 植被退化

在山前一带修建水库、拦蓄河水，平原地区大力开发地下水，由于忽视了生态用水需求，导致植被退化日益严重。

与此同时，灌区内因大水漫灌，地下水位不断抬升，次生盐渍化土地面积逐年增多，而水源地附近则由于抽水过量，地下水位持续下降，导致土地沙化。

3.3.2 土地盐渍化

土地盐渍化又叫土壤盐碱化，是指土壤层中可溶盐含量向增高的方向发展，成为盐渍化土的过程。一般将土壤层 0.2 m 厚度内可溶盐含量大于 0.1%的土壤称为盐渍土。

潜水埋深小于潜水临界埋深的细土平原，土壤及包气带蒸发失水，可通过毛细带源源不断地得到潜水水量和盐分补给，持续地蒸发可使土壤不断积盐。

4 山东省地下水可持续利用对策

地下水资源可持续利用的出发点和根本目的就是要保证水资源的永续、合理和健康的使用。一切与水有关的开发、利用、治理、配置、节约和保护都是为了使水资源在促进社会、经济和环境发展中发挥应有的作用。

山东省地下水分布广、储量大、调蓄能力强、水量水质相对稳定、保证程度高。地下水的开发利用，需要按照科学发展观的要求，制定实施地下水开发利用对策。

4.1 建设地下水库，实施地下调蓄

山东半岛年际降水量丰枯悬殊、河流源短流急、地表水与地下水相互间转化关系速度快,地下水所具有的年内和年际调蓄的功能对山东半岛地下水开发利用具有重要价值。

山东半岛山前冲洪积平原和山间河谷冲积层是一个巨大的天然地下水库，且具有埋藏浅、易采易补的特点。在枯水年份特别是在连续枯水年份，地表水源濒于枯竭、地下水天然补给量大幅度减少的情况下，合理开采利用地下水储存资源是解决无客水资源的山东半岛供水危机的唯一出路。

山东半岛地区地下水与地表水水力联系密切，如果以地表水调蓄为主，在河流上游大量修建地表水利工程，拦蓄地表水，势必改变地下水补给条件，造成山前冲洪积扇和河谷两岸地段地下水补给量大大减少，形成难以填平的地下水降落漏斗。建立地下水库，对水资源进行地下调蓄，与地表调蓄相比具有明显的优越性。根据水文地质条件分析，预计山东半岛可建设有坝地下水库和无坝地下水库 54 座，水库总面积 7 900 km^2，总调节水量 37 亿 m^3，可增加复蓄调节开采量 15 亿 m^3/a。

在海水(咸水)入侵多发区的山谷河流下游入海口处，以及山前冲积平原、冲洪积扇前缘建设地下水库，将大幅度提高地下淡水水位，增大地表径流量，恢复陆地地下淡水和海水(咸水)两大动力对抗中的平衡状态，这是治理海水(咸水)入侵的主要措施。

调蓄地下水资源、建设地下水库可以恢复水环境自然循环状态，增大环境容量，发挥地下水的自净能力，从根本上治理山东半岛目前较严重的水环境污染问题。

(1)典型实例：青岛大沽河地下水库。

青岛大沽河地下水库是青岛大沽河中、下游冲积层形成的水源地，砂层分布面积 293.24 km^2，砂层平均厚度 6.16 m，总体积 18.06 亿 m^3，其中可储水体积为 3.09 亿 m^3，地下水库相当于 3 个 1 亿 m^3 的大型地表水库的总库容量。

(2)典型实例：莱州王河地下水库。

莱州王河地下水库位于莱州湾南岸，地下水库设计库容为 5 693 万 m^3，最大调节库容 3 273 万 m^3。

王河干流长度 50 km，流域面积 326.8 km^2，地下水库库区总面积 68.49 km^2，干流长 10 km。1977 年以来，莱州市干流三次连续四年的严重干旱，全市掀起了打井高潮，沿海为 10 眼/km^2 以上，王河下游近 20 眼/km^2。过度超采导致海水入侵，1979 ~ 1999 年，海水入侵平均推进速度为 202 m/a，面积迅猛扩展至 278 km^2。王河下游是海水入侵的重灾区，近 800 眼机井变咸报废，6 万亩(1 亩=1/15 hm^2，下同)高产田变成低产田，耕地荒芜约 8 000 亩，地下水水质变差。水质恶化引发和加剧了某些地方病，发病人数近 3 000。为彻底改变海水入侵状况，莱州市利用日本协力基金贷款 6 230 万元，地方配套资金 8 956 万元，于 1999 年 12 月开工建设王河地下水库，2004 年 8 月竣工。王河地下水库利用 1 210 眼渗井和 187 条人工渗渠年回灌补源 1 180 万 m^3，利用人工湖调水年回灌补源 350 万 m^3，利用拦河闸汛期向洼台田沟引水，年回灌补源 300 万 m^3。

地下水库生态和环境效益巨大，主要表现为：

(1)提高了地下水位，减少了海水入侵面积。建库后比建库前地下水位最多提高 4.44 m，平均提高 3.31 m。

(2)改善了库区地下水水质，增加了供水能力。建库后比建库前氯离子含量最大减少 71.1%，平均减少 50.6%。

(3)增强了生态功能，减轻了自然灾害。王河地下水库能够储存来自大气降水、地下径流中的过多水量 3 273 万 m^3，恢复湿地 5 000 亩，避免了因季节和年度降水不均带来的洪旱灾害。

4.2 实施增源增采，涵养地下水源

充分利用鲁中南地区碳酸盐岩含水层分布广、储存空间大、调控能力强、入渗条件好、补给水源足的特点和优势，合理调控地下水位，增加地下储水空间，提高大气降水的有效入渗量，实现增源增采。

鲁中南地区寒武–奥陶纪碳酸盐岩分布面积约 2 万 km^2，是中国北方岩溶水资源分布面积最大的地区之一。区内发育有泰莱盆地、肥城盆地、临朐盆地、沂源盆地、陶枣盆地、羊庄盆地等。在许多中山、低山形成的单斜构造下缘和盆地第四系、新第三系下伏的元古代、古生代碳酸盐岩地层中，断裂构造、裂隙、岩溶均较发育，具有赋存裂隙岩溶水的优越条件，可形成多个大中型，甚至特大型集中供水水源地。

(1)岩溶水调蓄库容巨大。鲁中南地区寒武–奥陶纪石灰岩分布面积大，厚度达千余米，岩性稳定，分布广泛。受地质构造、岩性的控制，在山前及山间盆地、谷地中的隐伏灰岩分布区，岩溶裂隙较发育，形成岩溶水良好的储存空间，具有独立的补、径、排系统。这些岩溶水系统静态储存库容巨大，其丰、枯期的调蓄库容远远大于同类地区地

面水库容量。例如邹城市双村岩溶水水源地，隐伏灰岩分布面积 306.5 km^2，岩溶发育率为 20%，强发育带则大于 30%，岩溶地层综合储水系数为 0.072，每米水头高度的储水量达 398 万 m^3。

鲁中南地区岩溶水系统在区域水动力条件作用下，含水层各组段间具有相对统一的地下水位，因而具备作为统一含水层进行整体调蓄的基本条件。

(2)地表入渗条件优越。鲁中南地区碳酸盐岩裸露区岩溶洼地、岩溶漏斗、落水洞等岩溶地貌形态发育，为大气降水入渗补给与河流渗漏补给岩溶水创造了十分有利的条件。此外，境内修建的大量水库都不同程度地存在渗漏现象，从而成为现成的人工补给工程。

(3)补给水源较充足。本区多年平均降水量 750 mm 左右，降水量年际变化大，年内分配不均。降水多集中于每年的 7~9 月，占全年降水量的 60%~70%，这部分降水量在一些流域形成地表径流，是进行人工调蓄、增源增采的主要水源。

(4)人工补源工程形式多。鲁中南地区人工补源形式包括修建拦洪引渗工程、利用岩溶地貌形态人工补给工程、钻孔与坑道直接补给等。

典型实例：羊庄盆地岩溶水系统。

羊庄盆地岩溶水系统面积 650 km^2。地形上是一个四周环山的大型山间构造盆地，多年平均降水量 804.6 mm，6~9 月降水量占全年降水量的 73.8%，区内主要地表河流干流长 62 km，汇水面积 650 km^2，多年平均径流总量为 1.41 亿 m^3。

根据计算，在天然状态下，羊庄盆地岩溶水开采资源量为 24.02 万 m^3/d。采取涵养水源、增源增采措施后，岩溶水开采资源量增大至 32.93 万 m^3/d，较天然状态下增加 37%。

4.3　浅深开采适度，咸淡利用结合

全省大部分平原地区以开发浅层地下水为主，在沿黄两侧、山前冲洪积扇区深层地下水资源丰富而开发利用后又不产生较大环境地质问题的地区，可有计划地适度开发深层地下水；在东营、滨州地下咸水分布区可应用抽咸补淡、淡咸混合等技术合理利用咸水资源。

济南泉域回灌补源潜力研究

邢立亭，刘莉，张建芝，周瑞

(济南大学，济南，250002)

摘　要：本文分析了济南泉水动态特征,划分了济南泉域岩溶水系统“三水”转化的模式，找出了泉水断流的根本原因，进而提出了利用雨洪水直接补给区补源、控制城区扩展、优化开采布局等泉水保护措施。

关键词：“三水”转化；回灌试验；岩溶水；潜力；济南泉域

0　引言

济南以泉水名扬天下，在20世纪五六十年代，济南市区地下水年平均水位为32 m左右，趵突泉等四大泉群争相喷涌，景色壮观。随着济南市社会经济的发展，地下水位下降，泉流量减少。近几十年来，许多专家、学者对济南泉水进行了研究，主要集中在以下几方面：①济南泉域岩溶水化学成分的形成、演变趋势及防污染性能评价[1-4]；②水库放水试验介绍[5, 6]；③泉域边界条件的讨论[7, 8]；④限采、节水与水源涵养[9, 10]；⑤泉水动态监测与动态特征分析[11, 12]；⑥完善保泉的相关法规[13]；⑦泉域岩溶水数值模拟[14-16]等方面。综观以往研究成果，对于岩溶水资源构成及其水资源量演化的研究相对较少，本文从济南泉域“三水”(大气降水、地表水、地下水)转化关系分析入手，分析回灌保泉的可行性，供大家讨论。

1　泉水动态特征

据长期观测资料，济南泉水动态变化可分为以下5个阶段：1959~1967年高水位大流量阶段；1968~1975年中等水位中流量阶段；1975~1981年水位下降泉流量衰减阶段；1982~2002年低水位断续出流阶段；2003~2008年人工调控阶段。

从图1中可以看出，为保护名泉，自2003年以来实施关停自备井、减少水源地开采量等人工调控措施，加上气象因素的影响，2003~2008年泉水位一直维持在较高水平，与1968~1975年基本持平，但该时间段市区地下水水源地基本停采，多年平均降水量差别不大，但是泉水位、泉水流量均小于20世纪五六十年代，说明30年来地下水的补给条件发生了根本的变化。

2　济南泉域“三水”转化条件的演化

2.1　“三水”转化模式

济南泉域地处鲁中山地的北缘，地形南高北低。多年平均降水量为646.55 mm，古

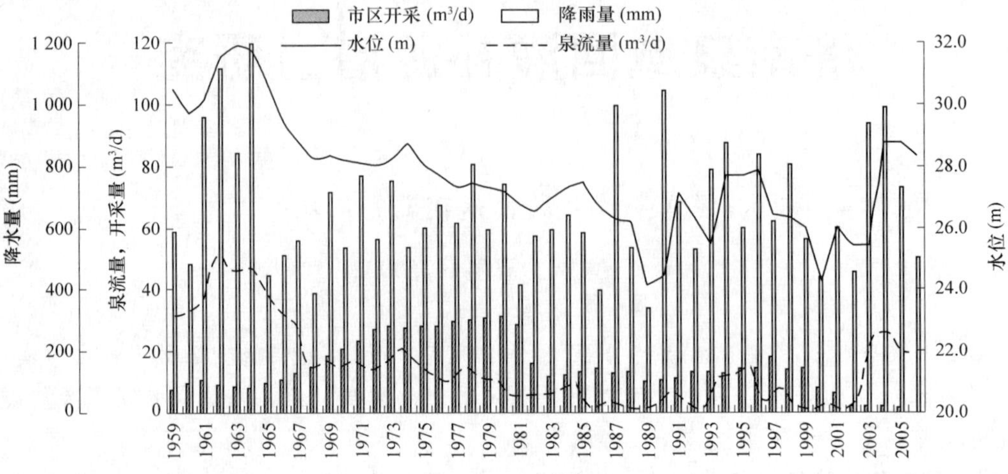

图1 市区地下水位与降水量和市区开采量对比曲线

生界寒武–奥陶纪碳酸盐岩裂隙岩溶含水岩组是济南泉域具有集中供水意义的含水岩组[17]。根据济南泉域地质、水文地质条件，按照地下水循环途径，将济南泉域地下水的转化关系划分为大气降水直接转化为地下水、地下水转化为地表水、地表水转化为地下水和地下水之间转化四种方式。概化济南泉域“三水”转化模式网络框图如图 2 所示，从图 2 中可以看出，济南泉水的形成极其复杂，控制岩溶水系统的水均衡方程为

$$(Q_1+Q_2+Q_3+Q_4+Q_5+Q_6)-(Q_7+Q_8+Q_9+Q_{10})=\mu F\Delta h \tag{1}$$

式中　Q_1——直接补给区降水入渗量；

Q_2——间接补给区地下水转化量；

Q_3——间接补给区水库灌溉回归量；

Q_4——直接补给区河水、沟谷渗漏量；

Q_5——孔隙水补给量；

Q_6——直接补给区井灌回归量；

Q_7——泉流量；

Q_8——人工开采量；

Q_9——转化的孔隙水量；

Q_{10}——径流排泄量；

μ——储水系数或给水度；

F——面积；

Δh——水位变幅。

从水均衡方程可以看出，对泉水位有重要作用的源、汇项主要是降水入渗、人工开采和河水渗漏。因此，研究地下水补给来源是保泉的关键问题。

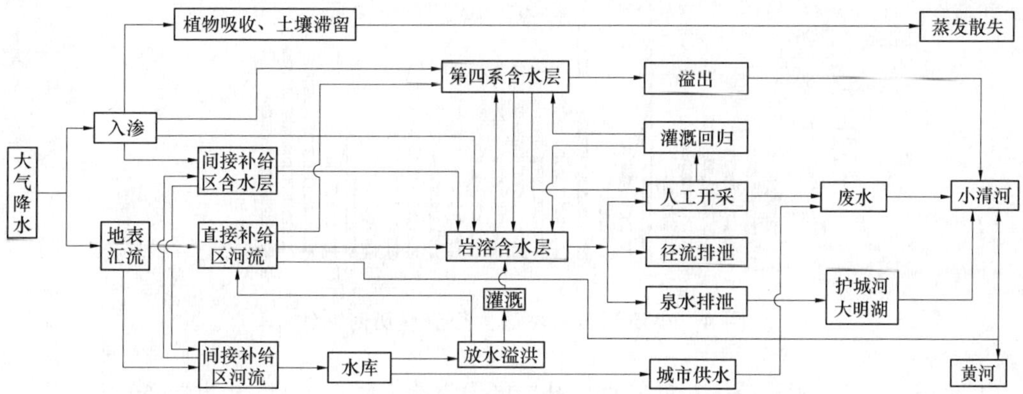

图 2　济南泉域“三水”转化网络框图

2.2　地下水补给区入渗功能退化对泉水补给的影响

2.2.1　来自间接补给区的河道径流补给大幅减小

济南泉域总面积 1 486.4 km^2，南部山区广大的间接补给区地表径流发育。根据“三水”转化关系，天然条件下，间接补给区表流通过河道进入直接补给区补给泉水。20 世纪 60 年代后，北沙河、玉符河、锦绣川等修建大中型水库，如卧虎山水库、岳庄水库等，拦蓄能力达 1.8 亿 m^3，自 1988 年卧虎山水库向市区供水以来，源自间接补给区的地表水补给逐渐减小(见图 3)，不能有效补给岩溶地下水。

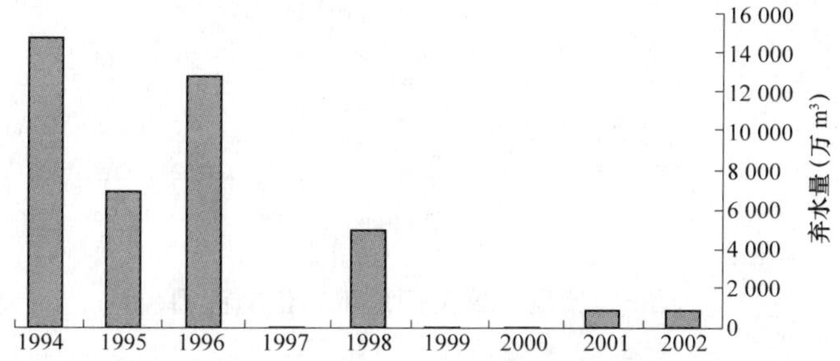

图 3　卧虎山水库弃水量柱形图

2.2.2　直接补给区河道断流、沟谷淤积，入渗能力降低

南部灰岩地区沟谷、河道是地下水重要的补给场所，自然条件下河流冲刷形成沉积韵律，结构松散，渗透性良好，人类活动对沟谷河道影响主要表现在河道占用、沟谷淤积、河道挖砂、沟谷垃圾堆积等方面。许多原本渗漏的河道、沟谷消失，严重淤积，河道涵养水源能力降低，造成泉水补给量减少。根据实测资料，目前玉符河河道单位长度的渗漏量不及 20 世纪 80 年代的 1/10。

2.2.3　城区扩展地面硬化，直接补给区面积减小、入渗能力降低

城区向南部扩展，直接补给区道路面积加大，地面硬化，地下水入渗量减少，20 世纪 60 年代市区径流系数 0.5，至 90 年代增加到 0.9[18]。由于建成区地面固化，地表径流量逐年增加(见图 4)，防洪压力增加。减少了地下水入渗补给，大部分降水大量流失。

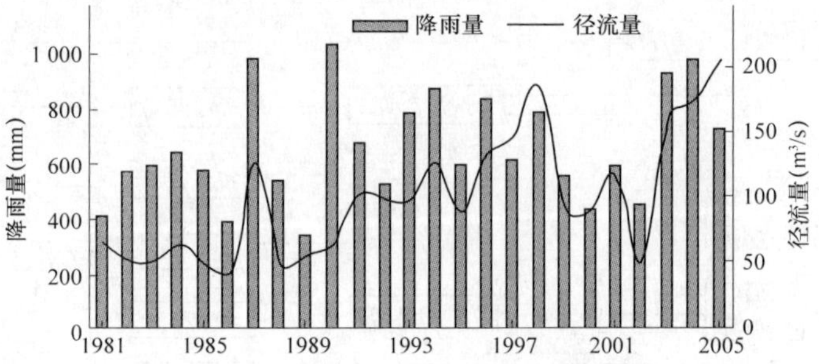

图 4　小清河黄台站最大径流量历时曲线

由于河道沟谷淤积、城区扩展使直接补给区入渗能力降低，与 20 世纪 60 年相比，区域地下水位总体下降(见图 5)，“三水”转化关系发生变化，一些转化方式消失，泉水补给量减少、丰齐—周王庄以北溢出量减少或消失，泉域地下水总资源量减少 35.6%(见图 6)，因此现状条件下即使减少地下水开采量，泉水位依然不可能恢复到 20 世纪五六十年代的高度，必须实施回灌补源。

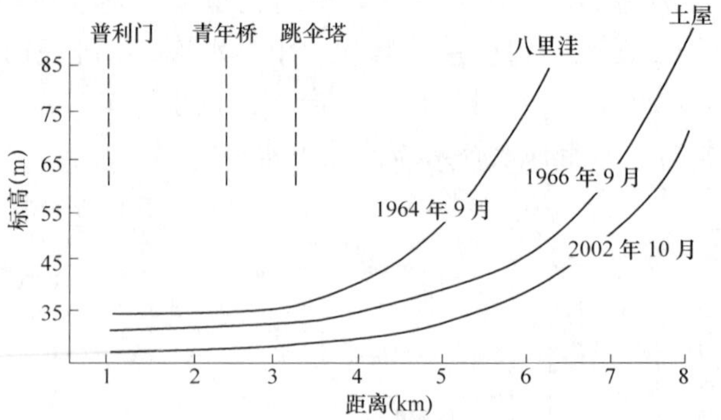

图 5　土屋—普利门地下水位剖面图

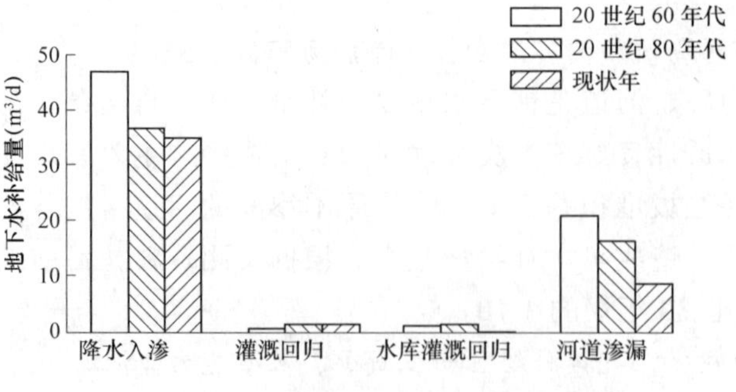

图 6　不同年代地下水资源量构成

3 保持泉水持续喷涌对策

3.1 回灌补源可行性研究

3.1.1 放水试验揭示的问题

根据 2003 年 3 月水库放水资料，绘制 3 月 17~18 日的回灌影响范围(见图 7)，地下水主流方向沿着炒米店地堑方向径流，西北方向的扩展速率较快，扩展范围较大，回灌期间地下水动态变化曲线(见图 8)显示西郊地下水位均有不同程度回升，回灌试验说明：

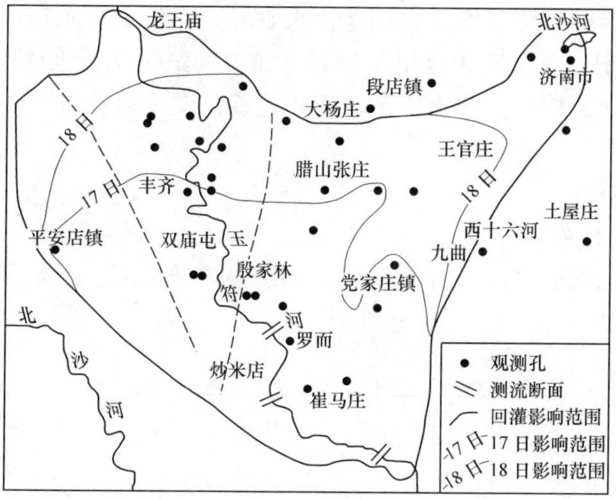

图 7 回灌试验影响范围示意图

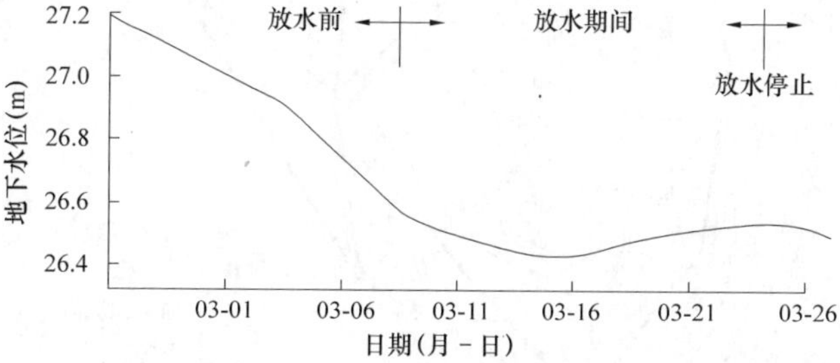

(a)市区水位变化曲线(2002 年 3 月 1 日～4 月 1 日)

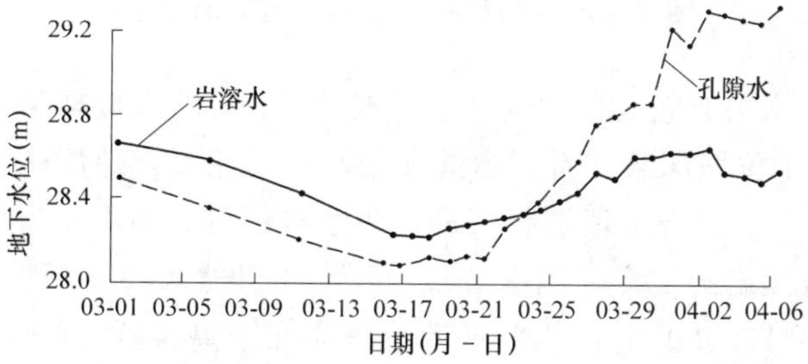

(b)西郊地下水动态变化曲线

图 8 回灌试验期间西郊地下水动态变化曲线

(1)玉符河回灌试验证实地表水是地下水的重要补给源之一；

(2)由于炒米店断裂的导水性和西郊开采影响，玉符河地表水首先填补水源地开采形成的漏斗，且影响速度较快，玉符河回灌对市区泉水的补给量相对有限；

(3)孔隙水与岩溶水动态对比表明，西郊孔隙水与岩溶水存在水力联系，大部分地表水补给第四系孔隙水，说明设计回灌试验 53.3 万 m^3/d 不合理，水库放水回灌补源量可设计在 18 万 m^3/d 以下。

3.1.2 示踪试验揭示问题

据连通试验，示踪试验各取样点初峰值来临时间，绘制含水层内示踪剂扩散速度等时线图(见图 9)。可见，不论在张夏组灰岩还是在奥陶纪灰岩含水层中，示踪剂主要运移途径沿北西方向扩散，指向西郊水厂开采区，示踪试验揭示张夏组灰岩与奥陶纪石灰岩关系密切。

3.2 回灌补源潜力分析

3.2.1 调蓄补源范围的确定

补源地段主要是依据岩溶发育程度、含水层的渗透性能、示踪剂运移途径、水位变幅等几个方面来确定。

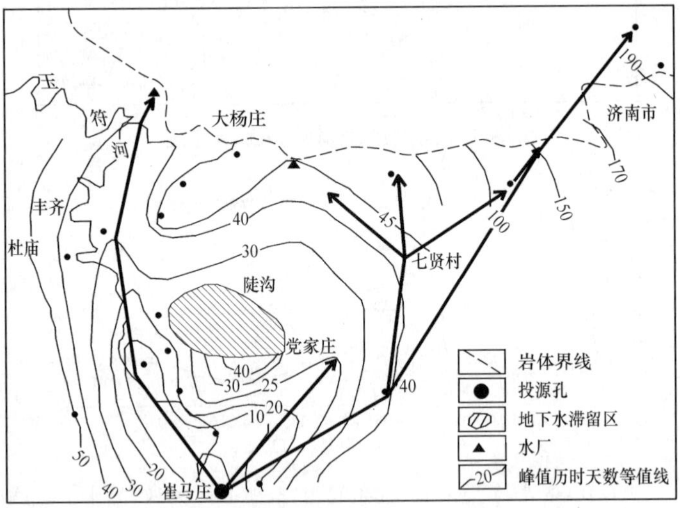

图 9 示踪剂初峰值扩散速度等时线图

济南泉域岩溶含水层有较大储水空间，根据计算，泉域直接补给区年水位变幅带储存量达 1.29 亿 m^3，从水位埋深条件看，山前地带水位埋深大，地层产状平缓，为北倾单斜构造，有利于自然入渗，补源地段选择山前地下水径流区，而不是市区附近，市区附近地下水具有承压性，虽然岩溶发育，但水位浅，不利于自然入渗，不宜净化补给水源。

按照一级保护区 100 d 的流程线，补源位置应在距离泉群 3.6 km 以南范围。综合野外实地调查分析，主要补源区为地下水的补给径流区，西起马山断裂，东至东坞断裂，面积 232.25 km^2，泉域内主要补源位置见表 1。

表 1 渗漏段基本情况表

位置	地层岩性	位置	地层岩性
北沙河	$€_2Z$、$€_3f-O_{1m}^2$	兴隆—土屋	O_{1m}^2
玉符河	$€_2Z$、$€_3f-O_{1m}^2$	扳倒井—土屋	$€_3f-O_{1m}^1$
锦绣川水库—西营	$€_2Z$	小岭子	$€_3f-O_{1L}$
小白庄—腊山	$€_2Z$、$€_3f-O_{1m}^2$	花峪—十六里河	O_{1m}^2
左而庄—二仙	$€_2Z$	柏石峪—十六里河	$O_{1L}-O_{1m}^2$
凤凰岭—韩家庄	$€_2Z$	中井—经十东路	$O_{1L}-O_{1m}^2$
泉泸水库—东家泸	$€_2Z$	浆水泉—经十东路	$€_3f-O_{1m}^2$
郭家窝—泉泸水库	$€_2Z$	南胡—刘志远	O_{1m}^2
铲村	$€_2Z$	燕棚—东梧	O_{1L}
义和庄—侯家庄	$€_3f-O_{1L}$	龙洞—石河岭	$€_3f-O_{1L}$

3.2.2 调蓄补源方式

针对济南地区降水特点和泉水补给条件的时空不均一性，实施人工回补地下水，从时间和空间上合理调配在枯水期用水量大而地下水补给不足的矛盾，人工补源是未来保泉的根本措施。补源必须从济南泉域的具体条件出发，充分考虑“三水”转化关系，以不同方式回补于地下含水层，进而从时空上调节水量以解决水量分配不均问题和控制地下水位。补源途径应考虑以下两方面：

一是直接补给区拦蓄地表径流，采用沟谷、塘坝入渗线状补给配合井点灌注，直接补给区补源应首先考虑洪水资源化[19]，济南城区洪水资源化可在保证安全的前提下，利用雨洪初峰清淤、在山前石灰岩沟谷拦蓄洪水或引水于枯水期放水渗漏补源，建设和完善雨洪利用体系可以起到防洪、补源、改善生态环境等多项功效。

二是采用卧虎山水库、岳庄水库在玉符河、北沙河补源，增加西郊开采优质地下水开采量。

4 结语

综上所述，人类活动改变泉域的水文地质条件，造成地下水资源量减少，增加了保泉的难度。正确地分析“三水”转化关系是研究泉水断流的关键，泉水恢复与保护必须与水文地质条件相结合，应注意以下两点：第一，回灌补源对保护名泉具有重要作用，但是回灌能力、回灌后对地下水水质影响尚无试验数据，需要进一步研究；第二，大气降水是地下水的主要补给来源，现状条件下控制城区扩展、优化开采布局是当务之急。

参考文献

[1] 万利勤,徐慧珍,殷秀兰,等.济南岩溶地下水化学成分的形成[J].水文地质工程地质, 2008, 35(3): 61-63.

[2] 徐慧珍,段秀铭,高赞东.济南泉域排泄区岩溶地下水水化学特征[J].水文地质工程地质, 2007,

34(3):15-19.

[3] 李常锁,胡爱民,游其军.济南泉域岩溶水水质演变趋势研究[J].山东国土资源, 2004,20(1):35-37.

[4] 邢立亭,康凤新.岩溶含水系统抗污染性能评价方法研究[J].环境科学学报,2007,27(3):501-506.

[5] 吴兴波,牛景涛,牛景霞.人工回灌对济南泉水系统修复的影响[J].人民黄河,2004,26(8):20-23.

[6] 牛景涛,吴兴波,宋星原.济南回灌补源与抽水试验研究[J].人民长江,2004,35(11):47-49.

[7] 李传谟,李岚,陶卫卫.济南保泉供水近期与长远对策[J].山东地质,2002,18(6):37-39.

[8] 商广宇,王明海.新水源开发与济南保泉供水[J].水文,2000,20(4):46-48.

[9] 李建江.济南泉水保护研究[J].水土保持研究,2003,10(3):142-145.

[10] 李望,刘济坤.济南市节水保泉实现水资源可持续利用的对策及建议[J].山东环境, 2002(2): 40-41.

[11] 刘国爱,赵新华.济南泉域岩溶水动态特征及有关问题讨论[J].山东地质,1997,13(2):67-69.

[12] 王庆兵,段秀铭,高赞东,等.济南岩溶泉域地下水位监测[J].水文地质工程地质,2007,34(2):1-7.

[13] 孟庆斌.济南“保泉”的制度分析和安排[J].中国水利,2003(8): 24-27.

[14] 张建国,陈鸿汉.济南泉域岩溶裂隙介质的多重指示克里格法研究[J].水文地质工程地质, 2004(2):25-28.

[15] 邢立亭.济南泉域岩溶地下水开发布局研究[J].人民黄河,2007,29(2):46-47.

[16] 武继虎.济南市泉水动态分析及数值模拟[D].南京：河海大学,2005.

[17] 奚德荫,李祥芝,邵卓,等.济南保泉供水勘探报告[R].济南：山东省地矿局 801 水文地质工程大队, 1988.

[18] 毛晓平.影响济南市区名泉停涌的主要因素分析[J].山东气象,2002,22(2):6-9.

[19] 向立云,魏智敏.洪水资源化–概念–途径与策略[J].水利发展研究,2005,(7):24-27.

王河地下水库水资源分析及工程建设状况

刘本华[1,2]，王维平[1]，初桂治[3]

(1 中国地质大学北京市重点实验室，北京，100083；2 济南大学，济南，250002；
3 山东省水利勘测设计院，济南，250002)

摘　要：本文根据地质勘查、野外试验所得到的数据，利用有限元法对王河地下水库区域进行水资源计算分析，并依据分析结果对地下水库的工程设计及施工工艺作简单介绍，工程完成后不仅能够防止由于地下水的过量开采所造成的海水入侵，同时将大量的降雨和地表径流转入地下水库中，能够满足该地区对水资源的需求，达到采补平衡、供需平衡的目的，防止环境地质灾害的发生。

关键词：地下水库；分区；负值漏斗区；工程建设

1　王河地下水库的意义及坝址区简介

莱州市是山东省工农业发展水平较高的地区之一，随着工农业的迅猛发展，用水需求不断增加，因大规模开采地下水，地下水位多年年均下降幅度达 1.0 ~ 1.5 m，加上王河上游修建拦蓄工程，使得滨海平原地区水位出现负值，随着时间的推移，已经形成了平行海岸线的三大串珠状漏斗。库区正位于三大漏斗区之一的市北漏斗区中心。负值漏斗区的形成加剧了海水入侵速度，导致地下水水质恶化，破坏了原有的生态环境，严重制约了工农业生产的发展，危及人民生活。

鉴于莱州市缺水的现实，从长远看，跨流域调水是解决该市水资源不足的最好途径，但因受各种条件的制约，短期内都难以实施。而在王河中下游修建地下水库，拦蓄地表径流补给地下水，充分开发利用王河流域水资源，发挥地下水库的多年调节作用，不仅能达到以丰补歉，增加地下水的储量和开采量的目的，而且也可有效控制与缓解区域内海水入侵的危害，是解决莱州市缺水、阻止海水进一步入侵的较为有效的途径。

王河地下水库工程是由日本投资的国家大型项目，该工程除修建地下北坝、西坝及副坝等挡水坝外，同时修建平里店、光明、院上及西由梯级拦水闸，在主河道及故河道内建补源渗井工程等。地下挡水坝北坝线西起仓上残丘东至街西残丘，全长 5 700 m；西坝北起仓上残丘，南至武家龙王河北岸，全长约 8 052 m；副坝为本次勘查新发现的一南北向隐伏故河道，位于西由镇新合村附近，宽度小于 850 m。

2　自然地理状况

拟建王河地下水库位于莱州市北 16 km，属王河中下游平原。西部、北部濒临渤海，东部以低山丘陵为界，南至龙王河，库区面积约 64 km^2，蓝三公路纵贯库区。

王河地下水库区域上属暖温带季风区大陆性气候，四季分明，冬季漫长干冷，降水稀少；春季干旱少雨，多春旱；秋季温凉；降水集中于夏季的6～9月，占全年降水量的74.5%，7月中旬至8月中旬为盛雨期，占全年降水量的53.4%。8月中旬以后，雨季结束。年降水量总趋势是自东南向西北递减。本区大气降水明显有年、季变化，雨季短且雨量集中，如图1所示。

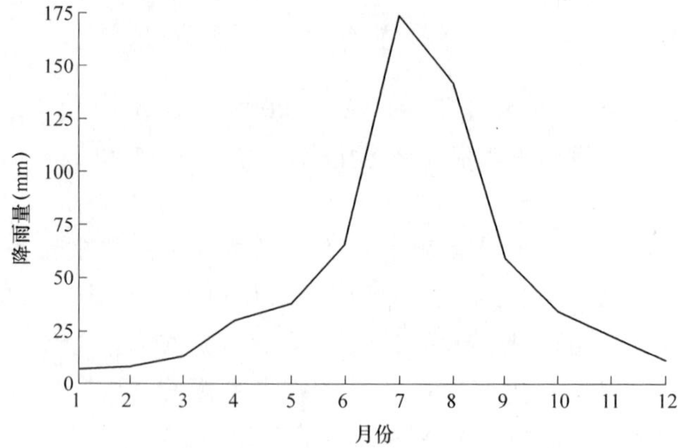

图1　逐年逐月(42年)平均降雨量过程线图

库区地势东高西低，从东南部低山丘陵至西北部沿海平原呈台阶式下降。王河发源于东南部低山丘陵地带，属于雨源型河流，且源短流急，流域面积326.8 km^2，全长50 km，在三山岛南入渤海湾，是莱州市规模最大的一条河流。

库区河川径流主要以大气降水为主，特点与大气降水基本一致，年际及季节性变化十分明显，再加上河道源短流急，汛期洪水暴涨暴落，在补给地下水之前，便大量注入渤海，枯水期则流量骤减，直至断流。

3　地下水库蓄水结构概述

3.1　边界条件

王河地下水库东倚山前岗地，西与渤海平原相接，东西宽 5～9 km，地形呈东南—西北方向带状展布，地貌上属冲洪积平原地带。地下水库主要储存介质即冲洪积堆积的粗砂、砾石、含土质粗砂砾石等富水性及渗透性强烈的介质。

地下水库的东部边界是北起街西村基岩裸露处，向东至桑家村南，由太古界、元古界胶东群民山组变质岩及燕山晚期侵入岩组成，透水性差，为相对隔水边界；西北边界南起仓上村南，北至仓上村北防护林，属仓上残丘，由燕山晚期花岗闪长岩组成，全风化透水率小于 38 Lu，属微透水层，作为相对隔水边界；北部边界东起街西村，西至仓上残丘，主要为第四系松散层堆积，厚度8.4～30.6 m，强透水，属透水边界。西部边界北起仓上残丘，南至龙王河，为冲洪积平原与滨海平原交互带，透水性强，为强透水边界，地下挡水坝建成后为隔水边界。南部边界东起曲家村北，西至武家村南，堆积物为第四系全新统河床相砾质粗砂，透水性强烈，属强透水边界。

库区基底是由太古界、元古界胶东群组成的片麻岩，岩石易于风化成土状，透水性差，且起伏不平(见图 2)，因此库区基地可作为隔水底板的隔水边界。

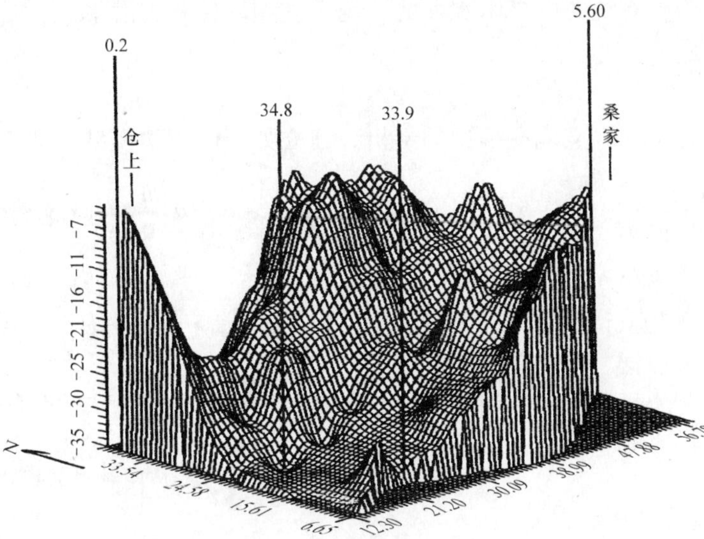

图 2　库区基岩埋深立体图

3.2　储水结构

库区含水层以第四系上更新统冲洪积堆积或洪冲积堆积成因为主，主要岩性为微含土砾质粗砂，局部相变为微含土砾质中砂、中粗砂等。在空间上分布有 3 ~ 4 层，总厚度一般为 5.0 ~ 15.0 m，自东向西逐渐变厚，最大埋深可达 23.7 m(见图 3)。含水层渗透系数一般为 16.9 ~ 89.3 m/d，少量超过 150 m/d，属较强—强透水层。

单井涌水量一般超过 500 m^3/d，少量超过 1 500 m^3/d，透水性及容水性较好。各层砂纵横方向连续性较好，各砂层间局部也能上下串通，水力联系较为密切，有利于各含水层间的相互补给，有利于地表水向地下水的转化，储水结构良好。

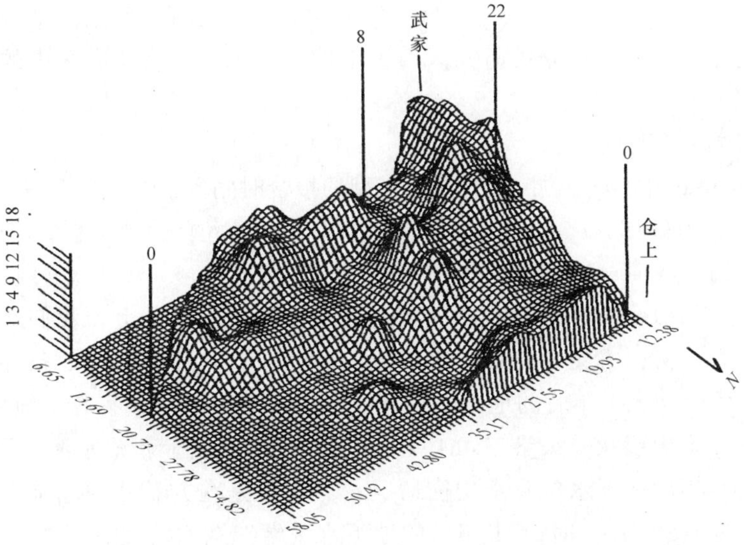

图 3　库区砂层厚度立体图

通过对王河地下水库边界条件及储水结构以及库区多年地下水流场的分析，可以将该地下水库概化为具有统一的自由水面的非均质各相同性三维潜水。

4 数学模型的建立

通过对王河地下水水文地质模型的剖析，建立如下的数学模型，即

$$\left.\begin{aligned}
&\frac{\partial}{\partial x}\left[k(h-b)\frac{\partial h}{\partial x}\right]+\frac{\partial}{\partial y}\left[k(h-b)\frac{\partial h}{\partial y}\right]+\frac{\partial}{\partial z}\left[k(h-b)\frac{\partial h}{\partial z}\right]+\omega=\mu\frac{\partial h}{\partial t} && \left((x,y,z)\in\overline{G},t>0\right)\\
&h(x,y,z,t)\big|_{t=0}=h_0(x,y) && \left((x,y,z)\in\overline{G}\right)\\
&h(x,y,z,t)\big|_{\Gamma_1}=h_1(x,y,t) && \left((x,y,z)\in\Gamma_1,t>0\right)\\
&k(h-b)\frac{\partial h}{\partial x}\cos(\eta,x)+k(h-b)\frac{\partial h}{\partial y}\cos(\eta,y)+k(h-b)\frac{\partial h}{\partial z}\cos(\eta,z)\Big|_{\Gamma_2} && =-q(x,y,z,t)\\
& && \left((x,y)\in\Gamma_2,t>0\right)
\end{aligned}\right\}$$

式中 η——水流的外法线方向；

Γ_1、Γ_2——计算区的第一类边界和第二类边界；

$\overline{G}$——计算区域；

ω——源汇项，$\omega=\varepsilon(x,y,z,t)-\sum_{i=1}^{v}Q_i\delta(x-x_i,y-y_i,z-z_i)$，其中 Q_i 为潜水含水层开采井的开采量，m^3/d，$\delta(x-x_i,y-y_i,z-z_i)$ 为 Delt 函数。

其余符号含义见相关的水文地质方面的书籍。

5 水资源计算及分析

5.1 计算区域的剖分及参数分区

根据计算区域内的勘查数据，将计算区域进行三角剖分，并对开采井点进行三角形的加密，共计三角单元数 1 788 个，结点总数 943 个，如图 4 所示。结合野外水文实验以及多年水文地质数据，在分析研究区内富水性的基础上，将研究区内的水文地质参数分为三层 36 个亚区，见表 1。

5.2 数学模型的模拟验证

选择 1990 年 6 月 1 日~9 月 15 日作为模型的拟合时间段；1990 年 9 月 15 日~1991 年 6 月 1 日为模型的验证时间段。通过计算结果与实际观测水位的对比分析，应用试估–校正法校核各参数区的参数值，如表 1、图 5 所示。

通过对模型的拟合验证，在对参数及边界条件校核的基础上，我们对该模型采用了三种方案的计算分析：方案 I 是地下水库建成后，在没有任何回灌的条件下，采用现状开采量 5.461 万 m^3/d，31 年后的地下水流场变化及均衡情况；方案 II 是地下水库建成后，在回灌条件下，采用现状开采量 5.461 万 m^3/d，并增加院上新建水源地开采量为 1.536 万 m^3/d 时，31 年后地下水位及水均衡情况；方案 III 是在方案 II 的基础上，新增尹家水源地的日开采量 1.50 万 m^3 时，31 年后的地下水位情况及水均衡状况。

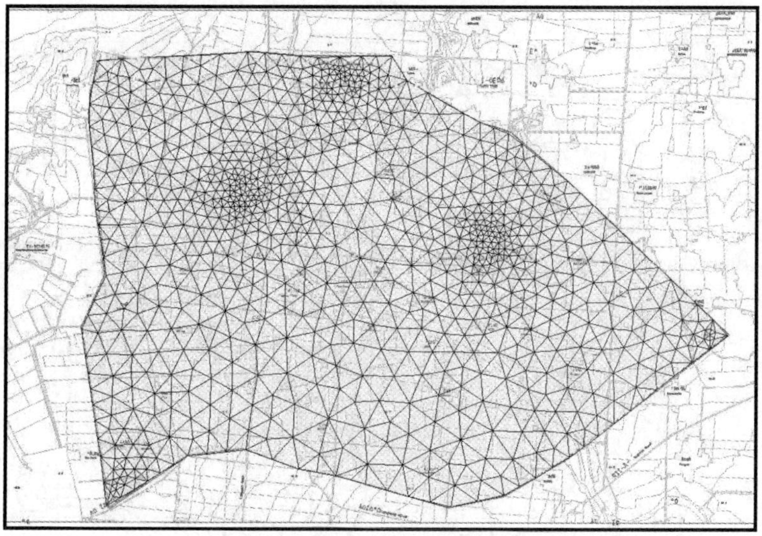

图 4　计算区域三角剖分图

表 1　计算区顶层分区参数表

参数分区编号	面积 (km^2)	渗透系数 k (m/d)	给水度 μ
Ⅰ	5.13	20.00	0.054 0
Ⅱ	22.72	45.00	0.134 0
Ⅲ	7.71	71.90	0.053 1
Ⅳ	2.91	27.00	0.178 9
Ⅴ	4.36	25.00	0.148 0
Ⅵ	2.67	55.00	0.108 0
Ⅶ	3.08	75.00	0.112 6
Ⅷ	5.54	60.00	0.000 3
Ⅸ	4.42	35.00	0.015 5
Ⅹ	0.30	30.60	0.127 4
Ⅺ	3.52	40.60	0.118 6
Ⅻ	4.19	89.70	0.160 0

5.3　计算结果及分析

5.3.1　方案Ⅰ

31 年后方案Ⅰ总的水均衡量为–8 400.00 m^3/d。表明地下水的补给不能满足现状开采，开采量大于补给量，地下水位将持续下降，地下水处于负均衡状态，海水将会持续不断地向大陆方向推进，不能有效防止海水的入侵。

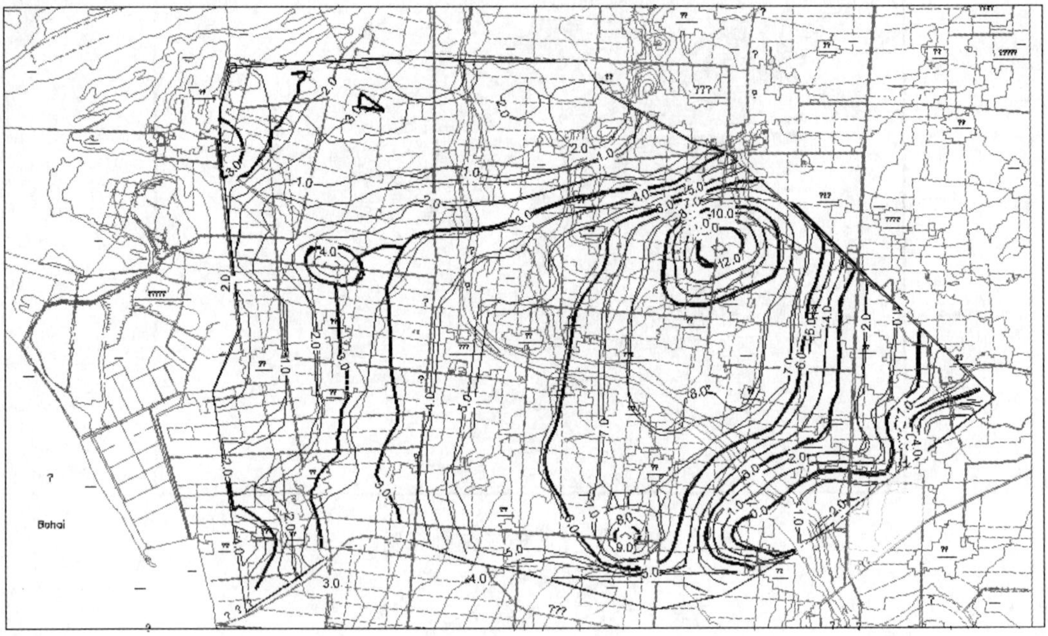

图 5　计算区域内拟合等水位线图

5.3.2　方案Ⅱ

地下水库建成后，在增加人工渗渠、渗井等回灌条件下，能够满足现状开采量 5.461 万 m^3/d 和院上水源地 1.536 万 m^3/d 的日开采量，31 年的水均衡量为 1.218 9 万 m^3/d，在满足开采的同时地下水位将会不断回升。

5.3.3　方案Ⅲ

在现状开采 5.461 万 m^3/d 的同时，满足院上水源地及尹家水源地各自开采 1.536 万 m^3/d 和 1.50 万 m^3/d，31 年后的总水量均衡为 110 m^3/d，表明补给量与开采量大体持平，旱季开采的地下水量在丰水季节能够补回，31 年内地下水的开采与补给基本平衡，如图 6 所示。

因此，通过计算分析可以看出，方案Ⅲ是最佳的工程方案。地下水库工程以该方案为基础进行设计。

6　地下水库工程建设概况

6.1　挡水坝址区岩性

根据水文地质及工程地质勘查可知，地下水库北坝、西坝及副坝坝址区主要岩性为第四系全新统冲积堆积及冲洪积堆积、海积堆积、上新统冲洪积的砂壤土、粉细砂、砾质粗砂、淤泥质粉细砂、壤土、砂质黏土等，底部为太古界—元古界胶东群民山组片麻岩以及燕山晚期的花岗闪长岩等。第四系堆积物渗透性强、厚度较大，需要进行工程防渗处理。不同坝线主要砂层厚度如表 2~表 4 所示。

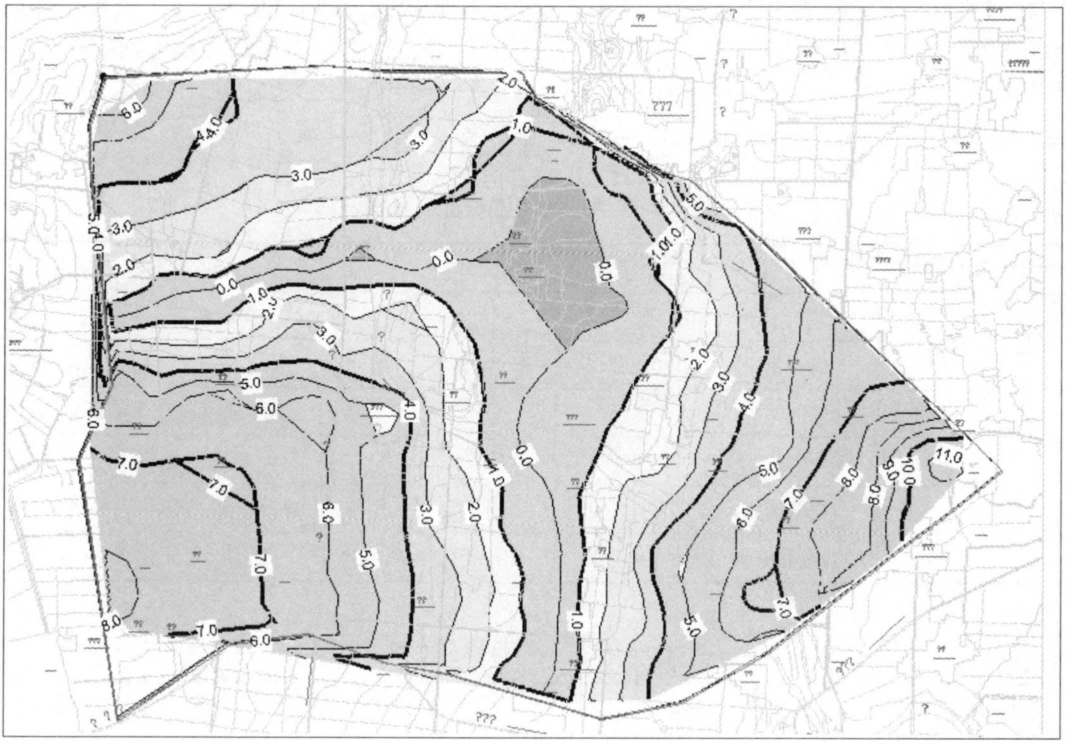

图 6 方案Ⅲ地下水库预测等水位线图

表 2 北坝址区岩性厚度一览表

时代	层号	岩性	厚度(m)	时代	层号	岩性	厚度(m)
Q_4^{al}	①–1	砂壤土	2.3	Q_3^{alp}	④–3	壤土	0.8 ~ 6.8
	①–2	粉细砂	2.2		④–4	含土砾质粗砂	1.1 ~ 5.5
	①–3	砾质粗砂	1.0 ~ 2.0	Q_3^{pal}	⑤–1	砂质黏土	3.1 ~ 5.5
Q_4^{apl}	②	砂壤土	1.5 ~ 2.5		⑤–2	微含土砾质粗砂	1.0 ~ 4.3
Q_4^{m}	③–1	砂壤土	0.7 ~ 2.5		⑤–3	微含土砾质粗砂	1.0 ~ 4.3
	③–2	砾质粗砂	3.3 ~ 8.6	$Ar-Pt_{1m}, \gamma\delta_5^{2(2)}$		基岩：变质岩及花岗岩	
	③–3	淤泥质粉细砂	1.2 ~ 6.7				

6.2 地下水库工程情况

6.2.1 主要工程

新建地下水库有地下挡水坝、西由拦河闸、地下回灌、过西橡胶坝以及过西引水渠等工程项目。其中，地下水库总库容 5 693 万 m^3，高喷板墙地下坝总成墙面积 30.42 万 m^2。西由拦河闸工程主要加固保留了老闸墩，并在原闸室上游新建 4 孔闸，闸室长 8 m，闸孔净宽 8.8 m，中墩厚 1.2 m，边墩 1.0 m。地下回灌工程为机渗井与人工渗井相结合，共打渗井 1 210 眼，渗井直径 900 mm，平均深度 17.5 m，梅花形布置，渗渠 65 条，底宽 2 m、长 80 m，平均深度 3 m。过西橡胶坝坝身结构型式为开敞式两孔橡胶坝，每孔净宽 64.5 m，坝身总宽 130 m，长 12.2 m，橡胶坝底板为分离式钢筋混凝土底板，橡胶坝组成部分主要有橡胶坝坝身、边墩刺墙，上游铺盖、护坦、下游消力池、海漫、防冲槽以及上下游翼墙和护砌等。过西引水渠工程全长 2.29 km，管道比降 1/1 000，设计渠底高程 5.4 ~ 2.25 m，渠顶高程 7.3 ~ 4.5 m，渠底宽 4 m，设计流量 7 m^3/s。

表 3　西坝址区岩性厚度一览表

时代	层号	岩性	厚度(m)	时代	层号	岩性	厚度(m)
Q_4^{al}	①–2	粉细砂	0 ~ 4.5	Q_3^{alp}	④–3	壤土	3.5 ~ 7.1
	①–3	砾质粗砂	0.8 ~ 6.4		④–4	含土砾质粗砂	1.8 ~ 5.9
Q_4^{apl}	②	砂壤土	1.5	Q_3^{pal}	⑤–1	砂质黏土	6.0 ~ 16.8
Q_4^{m}	③–1	砂壤土	2.2 ~ 2.4		⑤–2	微含土砾质粗砂	1.4 ~ 6.6
	③–2	砾质粗砂	1.4 ~ 4.3		⑤–3	微含土砾质粗砂	3.5 ~ 5.6
	③–3	淤泥质粉细砂	1.4 ~ 7.5	$Ar-Pt_{1m}, \gamma\delta_5^{2(2)}$		基岩：变质岩及花岗岩	

表 4　副坝址区岩性厚度一览表

时代	层号	岩性	厚度(m)	时代	岩性	厚度(m)
Q_3^{alp}	④–1	壤土	2.2 ~ 3.2	$Ar-Pt_{1m}, \gamma\delta_5^{2(2)}$	基岩：变质岩及花岗岩	
	④–2	粉细砂	1.4 ~ 2.6			
	④–3	微含土砾质粗砂	>6.1			

6.2.2　坝址区岩性

6.2.2.1　北坝

北坝西起仓上村北，东至街西村西北，该坝段位于滨海平原与山前冲洪积平原交互带，东西坝端分别与街西—后邓及仓上剥蚀残丘相接，基岩裸露。基底岩石为太古界—元古界胶东群民山组($Ar–Pt_{1m}$)老变质岩及燕山期花岗岩($\gamma\delta_5^{2(2)}$)，透水性微弱。第四系松散层岩性主要为砂壤土、细砂、粉细砂、砾质粗砂等，如工程地质剖面图(见附图 1)所示。

6.2.2.2　西坝

西坝南起朱由镇武家村龙王河北岸，北与仓上残丘相接，该坝段位于滨海平原与山前冲洪积平原交互带。第四系松散层为全新统冲积堆积层(Q_4^{al})、第四系全新统冲洪积堆积层(Q_4^{apl})、第四系全新统海积堆积层(Q_4^{m})、第四系上更新统冲洪积堆积层(Q_3^{apl})及第四系上更新统洪冲积堆积层(Q_3^{pal})五大层，见附图 2。

6.2.2.3　副坝

副坝位于西由镇新合村南，全长 840 m，根据钻孔及测绘调查，该处为一故河道垭口。据推测，该河道呈近南北向穿越新合村流入渤海。地层为第四系上更新统冲洪积堆积层(Q_3^{apl})，岩性组合为壤土、粉细砂、微含土砾质粗砂，底部基岩为太古界—元古界胶东群民山组片岩，揭露深度 9.7 m。

6.2.3 地下挡水坝坝型选择及比较

6.2.3.1 方案比较

地下坝多为机械施工，按构筑，地下坝施工方法可分为高喷板墙、混凝土防渗墙和垂直铺塑帷幕等型式，由于垂直铺塑施工深度有限，塑膜寿命短，不易在本坝址采用，本工程仅对高喷板墙和混凝土防渗墙方案作比较。

高压喷射灌浆工艺是利用钻机造孔，然后下灌浆管，其余土粒在喷射流的冲击力、离心力、重力等作用下，与灌入浆液掺搅、混合，形成板墙单体，多个单体按不同的结合型式构成一个完整的高喷板墙地下挡水结构，其主要优点是施工工序简单、造价低、适应变形能力较好，缺点是防渗板墙厚度不均匀。高压喷射灌浆技术在我国构筑水利工程防渗板墙中得到广泛应用，在山东省龙口八里河等地下水库也有成果应用的先例。

混凝土防渗墙(地连墙)在防渗工程中应用的历史悠久，它是利用钻机开槽孔，然后灌注混凝土或塑性混凝土。其优点是技术可靠、有效，并有现行施工规范可依；缺点是工效低、成本高、变形适应能力差。近年来，山东省太和水库、门楼水库的坝基截渗工程利用该项技术取得了理想效果。

1)方案一：高喷板墙方案

(1)墙厚：墙厚决定于板墙允许渗透坡降和使用年限。

高喷板墙凝结体为水泥砂浆和水泥土，其允许坡降和混凝土、土壤不同，山东省水利科学研究院在几个工地现场取样做渗透破坏试验，破坏坡降为 1/693 ~ 1/1 200，与混凝土允许坡降相比，考虑适当大一点的安全系数，允许坡降为 1/100，作用水头为 10 m，则板墙最小厚度选为 0.1 m。

板墙使用寿命不小于 30 年，影响其使用寿命的因素主要是混凝土中石灰淋蚀。一般将石灰淋蚀强度丧失 50%的历时作为使用年限。按阿达莫维奇公式计算，高喷板墙凝结体呈复合型，从内向外分为核心层、浆皮层和渗透凝结层，具有良好的抗石灰淋蚀性。工程实践表明，它与混凝土的使用寿命相当，经计算，高喷板墙厚 0.1 m，使用年限达 32.5 年以上。

(2)高喷形式确定。

高喷有定喷、摆喷和旋喷三种形式。在相同条件下，摆喷和旋喷凝结体的浆液成分含量比定喷小。其防渗性能定喷最强，摆喷次之，旋喷最差；有效厚度定喷最薄，摆喷次厚；旋喷为柱状。综合结构性能、防渗性能考虑，高喷板墙以墙高 10 m 分界，分界线以上部分采用定喷，以下部分采用摆喷，入基岩部分采用旋喷。副坝全部采用摆喷，入基岩采用旋喷；墙的连接，定喷和摆喷都采用折线式，喷射方向与轴线呈 20°，即连接角度 140°，摆喷的摆角 20°。

(3)材料的选择。

目前，高喷防渗墙一般采用纯水泥浆或水泥黏土浆。两种浆液形成的凝结体的物理力学性能有一定的差别。水泥浆在砾质粗砂中形成的凝结体类似于水泥砂浆，强度较高。水泥浆在黏土中和水泥黏土浆在砾质粗砂中形成的凝结体类似于水泥土，强度较低，但防渗性能较好，水泥浆和水泥黏土浆物理力学指标见表 5。

表 5 水泥浆和水泥黏土浆物理力学指标

浆液	水泥成分			
	水泥含量(%)	抗压强度(MPa)	渗透系数(cm/s)	弹性模量(MPa)
水泥	30 ~ 60	10 ~ 20	15^{-5} ~ 10^{-7}	10^{3} ~ 10^{4}
水泥：黏土(1：1)	20 ~ 30	3 ~ 5	10^{-5} ~ 10^{-7}	10^{2} ~ 10^{3}

根据承受水头不大的防渗墙，强度要求容易得到满足，根据经验，选用水泥浆液。主要工程量如表 6 所示。

表 6 高喷防渗墙主要工程量表

项目		单位	北坝		西坝	
			定喷	摆喷	定喷	摆喷
成墙面积(m^2)	总面积	m^2	93 923		164 685	
	砂层	m^2	27 294	10 766	17 060	48 070
	基岩	m^2		2 815		3 720
	其他	m^2	15 098	37 950	26 520	69 315
钻孔进尺(m)	钻孔进尺	m	65 668		110 556	
	砂层	m	26 718		45 578	
	基岩	m	1 759		2 325	
	其他	m	37 191		62 653	

2)方案二：混凝土防渗墙

(1)造孔设备。

造孔设备有两类：一类是直接出碴的挖(抓)斗式挖槽机，一类是泥浆循环出碴的钻头式挖槽机。目前，国内常用液压抓斗和冲击钻，液压抓斗成孔费用低、功率高，本工程拟定用导板式液压抓斗，型号为 BS655 型，主要指标如表 7 所示。

表 7 BS655 型板式液压抓斗指标表

指标	性能	指标	性能
主卷扬启闭力(t)	50	斗重(t)	13
副卷扬启闭力(t)	15	最大挖深(m)	50
斗体积(m^3)	0.5	导板长(m)	12
斗宽(m)	0.8	不垂直度(%)	0.6
开斗长(m)	2.8	适合挖掘土类(m)	砂砾土及风化岩

(2)墙厚。

墙厚主要决定于成槽机械，采用导板式液压抓斗，墙厚为 0.8 m，可满足结构抗渗和耐久的要求。

(3)单元槽段长度。

槽段长度决定于工程地质情况、成孔深度和固壁泥浆质量，在保证槽壁稳定的情况下，槽段越长越好，尽量减少槽段接头。泥浆用 4：1 的膨润土和黏土配制泥浆，其性能如表 8 所示，其主要工程量如表 9 所示。

表 8 泥浆性能指标表

黏度 (Pa·s)	比重	含砂量 (%)	胶体率 (%)	稳定性	失水量 (mm/30 min)	静切力 (mm/cm^2)		泥饼厚 (mm)	pH 值
						1 min	10 min		
18～25	1.1～1.2	≤5	≥96	≤0.03	20～30	20～30	50～100	2～4	7～9

表 9 北坝及西坝混凝土防渗墙工程量一览表

项目		坝段	
		北坝	西坝
成墙面积(m^2)		93 923	164 685
槽段个数		801	1 150
导板式液压抓斗进尺(m)	砂层	13 593	23 261
	基岩	1 005	1 329
	其他	18 946	34 227
200#混凝土导墙方量(m^3)		7 739	11 112

6.2.3.2 地下坝方案分析

上述就高喷板墙和混凝土防渗墙各自的特点、厚度确定、材料选择、施工工艺及施工参数等方面作了分析比较，下面从防渗效果和工程投资作比较。

1)防渗效果分析

在国内防渗墙的使用中，墙高在 30 m 以内且地质条件较好的地区，大都采用高喷防渗墙，防渗效果一般较好；超过 30 m 的防渗墙大都采用混凝土防渗墙，防渗效果较好。山东省龙口市地下水库地下坝虽有部分防渗墙超过 30 m，最深处达 40 m，但因合理选择工艺参数、严格控制施工质量，高喷防渗墙效果依然满足要求。

2)工程投资分析

高喷防渗墙工程和混凝土防渗墙工程可比较投资分别为 5 603 万元和 19 654 万元。

6.2.4 方案选定

高压喷射防渗墙工程在我国相同地质情况下的地下水库建设中已有成功的先例，在仓上金矿矿坑防渗工程中得到成功的应用，且投资少、施工方便。经全面分析，王河地下水库选用高压喷射防渗墙作为地下坝的结构型式。

王河地下水库工程除地下挡水坝工程外，还有人工补源工程、泵站、净水厂、西由拦河闸等，在此不再一一叙述。

7 地下挡水坝工程设计

7.1 坝顶高程及坝基处理

北坝坝顶高程 2.0 m，西坝坝顶高程 1.0 m，副坝坝顶高程 6.0 m，坝轴线处的坝顶高程均低于地面 2.0 m，地下坝底部均插入基岩 0.5 m，采用摆喷、定喷工艺直至坝顶。

7.2 结构设计

7.2.1 北坝

北坝西起过西镇仓上村北岩石出露处，向东至街西村西北，全长 5 608 m，最大墙高 34.13 m，设计墙顶高程 2.0 m，以墙高 10 m 为分界线，分界线以上采用定喷，以下采用摆喷，墙底插入基岩 0.5 m 采用旋喷孔距 1.6 m，采用 20°摆角，折线连接。

7.2.2 西坝

西坝南起朱由镇武家村龙王河北岸，北与仓上残丘相接，全长 8 052 m，最大墙高 36.81 m，设计墙顶高程 1.0 m，其他布置同北坝。

7.2.3 副坝

副坝位于西由镇新合村南，全长 840 m，设计墙顶高程 6.0 m，全部采用摆喷，孔距 1.6 m，摆角 20°，折线连接，底部插入基岩 0.5 m，采用旋喷，工艺参数见表 10。

表 10 高喷工艺参数一览表

项目	参数	单位	摆喷	定喷
高压水	压力	MPa	40	37
	流量	L/min	75	75
压缩气	压力	MPa	0.8	0.7
	流量	m^3/h	>100	>100
水泥浆	压力	MPa	0.2 ~ 0.6	0.2 ~ 0.3
	流量	L/min	>50	>75
	比重		1.6 ~ 1.7	1.6
提升速度		cm/min	6	8
摆动角度		(°)	20	0
摆速		r/min	8	0
孔距		m	1.6	1.6

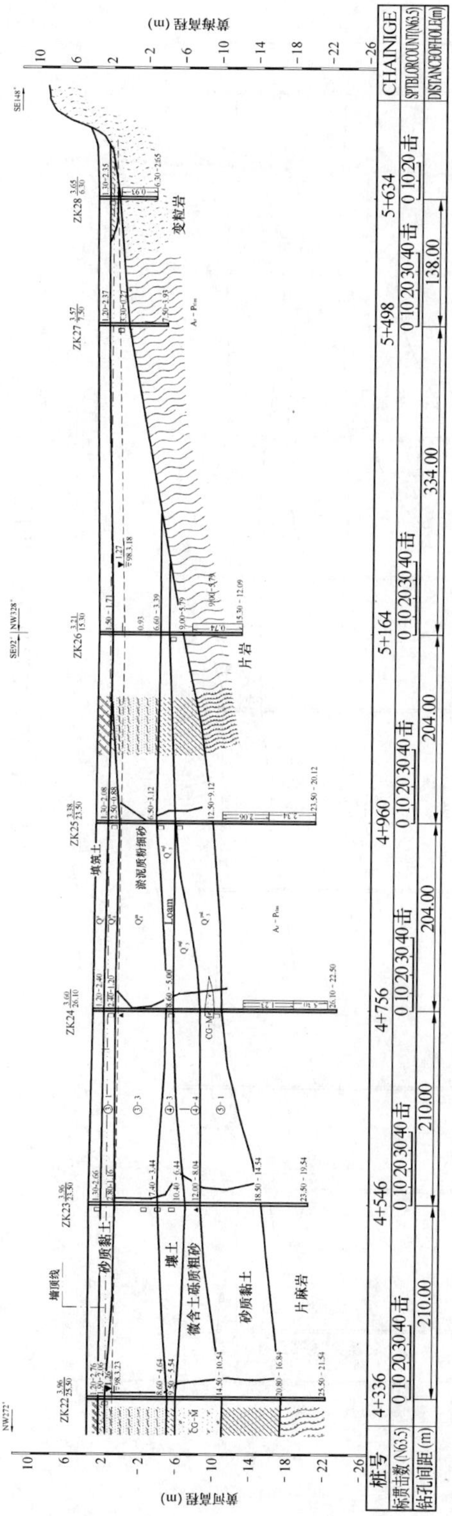

附图 1　挡水北坝工程地质剖面图

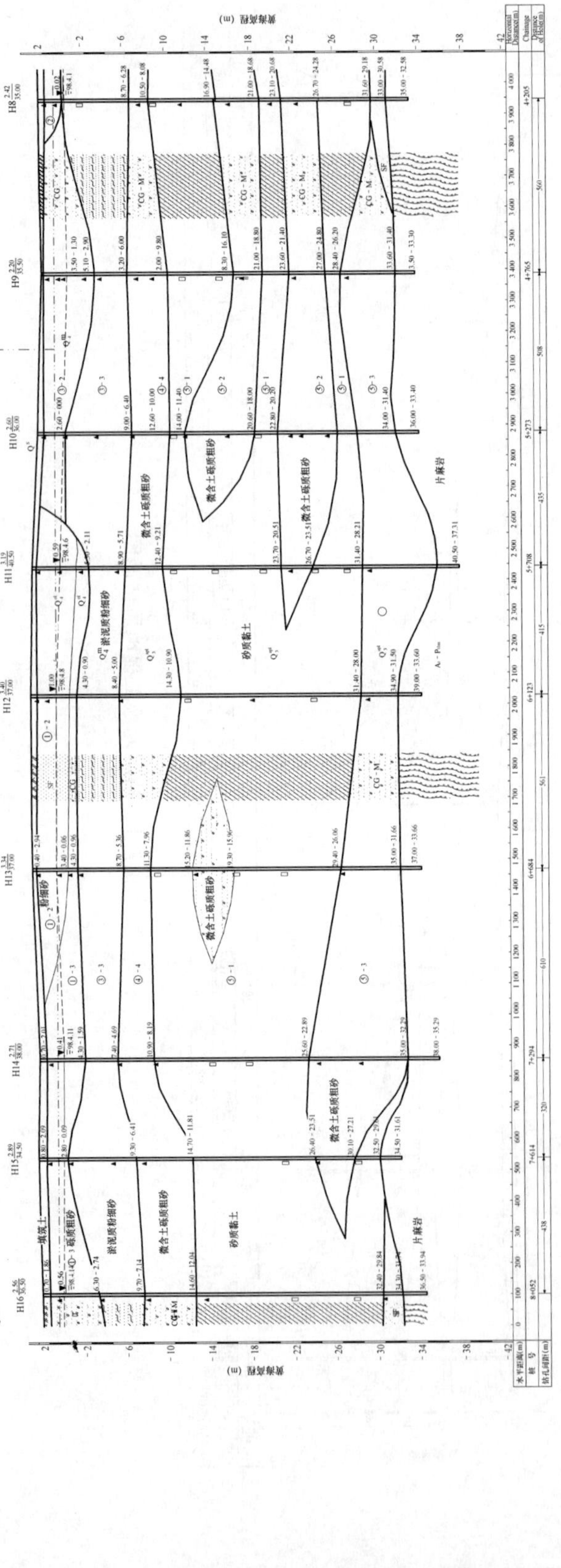

附图 2 挡水西坝工程地质剖面图

某炼油厂地下水系统石油烃污染物运移预测研究

张亮，马振民，于玮玮，李玲玲

(济南大学，济南，250022)

摘　要： 石油开采及其加工业致使大量石油烃类污染物进入到地下水系统中，破坏了原有的生态系统且危害人类的健康。本文选取某炼油厂地下水系统为研究对象，考虑机械弥散作用的影响，采用试验和数学模拟的方法建立了二维地下水系统石油烃污染物运移模型，对研究区内地下水系统石油烃污染物迁移进行了预测，预测结果表明2008年枯水期地下水系统石油烃污染面积为4.08 km^2，丰水期石油烃污染面积为4.56 km^2；2013年枯水期地下水系统石油烃污染面积为6.37 km^2，丰水期石油烃污染面积为7.17 km^2。

关键词： 地下水系统；石油烃污染物；运移；预测

在石油开采、炼制和储运过程中，由于生产流程不封闭，容器和管道破裂，直接导致大量石油类污染物进入土层中，对周围环境会产生多种污染和破坏。进入土层的污染物小部分直接进入含水层中，大部分长期残留在包气带土层中，通过地面水体及降水的不断淋滤和冲刷，致使油类污染物长期向地下水中释放，有些地区的地下水受到严重的污染[1,2]。石油烃有许多致癌、致畸和致突变的潜在性化学物质，进入人体后对人体的健康有很大的危害，因此石油烃污染地下水系统吸引了国内外许多学者的重视[3-13]。

1　研究区概况

某炼油厂位于某省会城市的东部，地下水系统所在区域位于单斜构造的翼端，研究区内发育的地层主要为奥陶系石灰岩、白云质石灰岩。地下水系统是由某炼油厂所在场地西侧、东南侧区域性断裂和北侧隐伏于第四系以下的东西向断裂构成的三角形断块，奥陶系石灰岩地层和断裂构造控制着研究区的水文地质条件。地下水系统主要含水岩组为奥陶系石灰岩、白云质石灰岩，含水层分布稳定，岩溶裂隙较发育，渗透性和导水性较强，单井涌水量500~1 000 m^3/d，富水性中等。

大气降水是地下水系统的主要补给来源。大气降水除形成短暂的地表水流外，部分沿石灰岩裂隙、岩溶渗入地下水系统，补给岩溶水含水层。区域性断裂对地下水的补给、径流及排泄起一定的控制作用。

2　地下水系统石油烃污染特征

石油是一种轻的非水溶性液体(Light Non-aqueous Phase Liquid)，简称LNAPL[14]，

它泄漏到地下水系统以后，在污染源附近浮于地下水表面，形成浮油，浮油向下移动非常缓慢。在浮油底部和四周，由于地下水的流动一部分油组分溶于地下水或形成乳化油并随地下水运移至远处。溶解到水中的石油烃既可随地下水做对流流动，也可以因水动力弥散迁移。

地下水系统石油烃污染物的迁移转化是综合物理化学及生物作用的过程，其中包括对流作用、分子扩散作用、吸附与解吸作用、放射性衰变、生物降解，以及吸附滞留。[15] 适应本研究区的真实情况，并考虑石油烃污染物预测模型的精确性和可用性，在建立预测模型的过程中只考虑上述因素中的个别条件。地下水系统石油烃污染物有两个明显的特点：具有隐蔽性，从地表很难察觉；极难治理，地下水系统一旦遭到污染，短期内很难恢复。

3 地下水系统石油烃污染物运移预测

某炼油厂地下水系统石油烃污染物运移预测主要分为两类：进入岩溶裂隙并浮于地下水表面浮油污染物的运移预测及岩溶裂隙含水层的溶解或溶混于地下水中石油烃污染物的运移预测。地下浮油污染物预测主要是通过确定阻隔浮油运移的隔油边界，以及隔油边界与地下水位的相关规律来实现的；地下水中石油烃污染物的预测，首先是通过现场试验求得各项环境水文地质参数，然后运用建立的污染运移模型进行模拟预测。

3.1 模拟预测域的选择及研究程度

某炼油厂所处的位置，地下水的流向、流速等特征限定了污染模拟区的范围。模拟区包括现有炼油厂的污染域，以及厂区以南、以北、以东和以西的广大区域。

因没有某炼油厂地下水系统石油烃污染预测区内多年连续的水质资料，本次研究采用 2002 年 12 月~2003 年 12 月的地下水水质资料，同时收集了本区域内的以往部分地质、水文地质勘探资料，为建立某炼油厂地下水系统石油烃污染物运移预测模型工作奠定了基础。另外，对区域地下水氚同位素的分析证明，岩溶水含水层主要接受研究区以南石灰岩裸露补给区域的地下水径流补给及大气降水补给。为了预测地下水石油烃污染物运移规律，在厂西及厂北钻孔中做了 ^{131}I 同位素弥散试验。通过试验初步掌握了区域地下水系统岩溶裂隙水的基本特征，并求得了地下水系统石油烃污染物预测模型中所需的各项环境水文地质参数。

3.2 地下水系统石油烃污染物运移特征

根据地下水系统岩溶裂隙中浮油污染分析，浮油在地下水自由表面上部的裂隙空间的运移主要取决于地下水面以上岩溶裂隙的连通性。

研究区域内普遍存在的红色黏土层且浅部岩溶裂隙中被红色黏土填充的裂隙充填层具有较好的阻油性能。在研究区北部随着第四系厚度的增大，红色黏土层的埋深也相应增加，使区域地下水呈现出微承压的特征。这样，裂隙充填黏土层就可作为限制地下水系统浮油污染物运移的可变边界。

在地下水系统岩溶裂隙中，浮油赋存于地下水的自由表面上部，地下水位上部的裂隙空间是浮油运移和储存的主要场所。根据这一原理，地下水位标高等值线与隔油层底板标高等值线的相交部位就是浮油运移的最大范围的边界，因此可按区域地下水位的变化来预测浮油污染的范围及污染的运移方向。当地下水位下降时，浮油的运移边界将向外推移，最大污染范围扩大。

2003 年地下水系统岩溶裂隙中浮油污染物运移监测结果为：丰水期浮油污染物面积 0.82 km^2，枯水期浮油污染物面积 1.06 km^2。分析表明：赋存于某炼油厂污染区的地下浮油在现有地下水面形态的条件下，将逐渐向西北方向运移。由于浮油运移范围不断加大，浮油层逐渐变薄，这时浮油层的黏度及岩溶裂隙的吸附作用等将会更有效地限制浮油的运移速度。在不存在新的浮油污染物进入地下水系统岩溶裂隙的条件下，浮油的运移速度将随污染范围的不断扩大而减小。

3.3 地下水系统石油烃污染模拟与预测

3.3.1 野外弥散试验及求解水文地质参数

3.3.1.1 试验选址

野外弥散试验的位置选在某炼油厂厂区北东方向 1 000 m 处，即污染域北部地下水流方向的尚未污染区，此处获得的水文地质参数更接近预测区的实际水文地质条件。试验场施工 3 个钻孔，其中 1 个主孔、2 个测孔，孔深均在 100 m 左右。3 个试验孔的主要含水段均位于 37~80 m 的两个层位，岩性为奥陶系石灰岩、白云质石灰岩。两个观测孔皆位于主孔的 NE 方向上，距离主孔均为 8 m，两观测孔间的距离为 2.97 m。观测孔位的方向与孔距是根据主孔中测定的地下水流向与渗透速度以及区域地下水流方向确定的。

3.3.1.2 试验步骤

(1)初步试验。为测试含水层在不同深度的渗透速度，分别在 40~50 m 处分段投 ^{131}I 源，测得钻孔内局部渗透速度(m/d)为

$$v_f=0.98\pm0.22$$

(2)初步连通性弥散试验。在主孔内深度 40 m 处投 ^{131}I 源，随后在主孔中注水，进行初步连通性弥散试验，1 h 后在观测孔中接收到了示踪剂进入观测井的伽马信号。

(3)稳定流注水弥散试验。在主孔内深度 40 m 处投放 ^{131}I 源，随后在主孔连续注水，平均注水量为 9 m^3/h，3.5 h 后接到了示踪剂的峰值。本次试验共连续注水 5 h，总注水量达到 45 m^3。试验所采用的方法基本上是模拟某炼油厂地下水系统的石油烃实际污染过程，属于稳定流注水弥散试验。

(4)天然流场的野外弥散试验。为了获得更准确的天然条件下地下水系统含水层的水文地质资料，在试验场又进行了地下水系统天然流场的野外弥散试验，在主孔内投放 ^{131}I 源后 9 h 接到了 ^{131}I 源的浓度峰值。

(5)环境水文地质参数计算。对弥散试验所获得的实测资料进行系统归纳整理后，应用上述公式求得了区域地下水系统岩溶裂隙含水层的弥散度以及其他参数如下：

区域地下水天然流速 v=13 m/d；

含水层的厚度与裂隙率之积 $B \cdot N$=0.118；

区域地下岩溶裂隙水的弥散度 a_l=1.6。

以上所求得的各项环境水文地质参数是在尚未被石油烃污染的预测区域内，是通过模拟某炼油厂地下水系统的石油烃污染过程和进行多次对比弥散试验取得的，因而数据

比较可靠。

在计算稳定流注水试验资料时，选用了一维流场的拟合法求解纵向弥散度。其无量纲弥散公式为

$$C_R = K \Big/ t^{1/2} \cdot \exp\left[\left(-P/4t\right)\left(1-t\right)^2\right]$$

式中

$$K = t_p^{1/2} \exp\left[P \Big/ 4t_R\left(1-t_R\right)^2\right]$$

$$t_R = \left(1 + 1 \big/ P^2 - P^{-1}\right)^{1/2}$$

通过上式求得 P=5(当 $P>1$ 时，理论弥散曲线与实测弥散曲线拟合得非常理想)，纵向弥散度 $a_l=S/P=8/5=1.6$ m。

3.3.2　模拟污染物的选择

根据地下水环境调查资料，某炼油厂地下水系统污染域中较为普遍存在的污染物有：石油烃类、硫化物、NO_3^-等。石油烃类有机污染物在地下水中的浓度分布较为稳定，且有规律可循，污染域也大于其他污染物的影响域，是某炼油厂地下水系统的主要污染物质，所以选择地下水中石油烃污染物作为污染质。

选用石油烃类有机污染物建立地下水污染物运移模型，存在的主要问题是地下水系统中石油烃类有机污染物极易被岩层介质所吸附，从而减少其在地下水中的含量。消除地下水中易被吸附污染物的影响，可通过修正污染源注入边界的浓度值来解决。本文所选的石油烃类污染物的注入边界浓度值为 1.0 mg/L，这个数值是根据注入边界及地下水开采井中石油烃类浓度值所确定的。

3.3.3　地下水系统石油烃污染物运移预测方法

某炼油厂地下水系统污染域及预测域发育为岩溶裂隙含水层，属于典型的非均质介质。因石灰岩中存在弥散作用且主要为水动力扩散，所以现有的可渗透触媒中水动力扩散研究结果可用于裂隙介质。分析某炼油厂地下水系统奥陶系地层的石灰岩裂隙可知，模拟区主要区段的岩溶裂隙含水层的裂隙、岩溶具有密集裂隙介质、层位稳定的特征，且该层是某炼油厂石油烃污染物运移扩散的主要层位，因此可概化为均质介质特征并以此为基础建立地下水污染物运移模型。

地下水系统石油烃污染物运移预测模型中除数值模型外，解析方程建立的数学模型也很有应用价值。解析方程要求较为理想的水文地质条件，生产实际中很难满足。如选用适当条件、可靠参数则用解析预测模型计算的精度与数值预测模型的预测结果相比其误差很小。

某炼油厂地下水系统的污染预测模型中所采用的各项参数主要依据预测区岩溶裂隙含水层中实际的野外试验数据求得。某炼油厂地下水系统污染预测区范围较小，用解析方程进行石油烃污染物运移预测，在现有资料及条件下能比较实际地反映地下水系统的石油烃污染物运移规律。

3.3.4　地下水系统石油烃污染预测模型

某炼油厂地下水系统石油烃污染物运移预测区的含水层可概化为等厚度无限边界。

从某炼油厂地下水系统的现状污染特征及污染域的展布形态分析，某炼油厂的地下水系统石油烃污染可视为南北两个不同径向流区分别接近于径向流弥散类型。

通过对某炼油厂地下水系统含水层的概化，考虑弥散作用等过程影响，建立二维石油烃污染质运移模型，即地下水系统石油烃污染物运移预测模型基本形式为

$$R_d\frac{\partial C}{\partial t}=\frac{\partial}{\partial x}\left[N\left(D_{xx}\frac{\partial C}{\partial x}+D_{xy}\frac{\partial C}{\partial y}-Cv\right)\right]+\frac{\partial}{\partial y}\left[N\left(D_{xy}\frac{\partial C}{\partial x}+D_{yy}\frac{\partial C}{\partial y}-Cv\right)\right]+GC_0-WC$$

式中 R_d——污染物在地质介质中的滞留因子；

D——弥散系数张量；

v——渗透速度；

G——补给项(如降雨入渗等)；

W——排泄项(如蒸发、抽水等)；

C——污染物浓度；

C_0——污染物初始浓度。

补充初始条件与边界条件，有

$$C(0,0,t)=C_0 \quad (t>0)$$

$$C(x,y,0)=0 \quad ((x,y)\in \Omega)$$

经计算，可求得石油烃类有机污染物浓度与扩散距离和时间的关系式为

$$C(x,y,t)=\frac{1}{2}C_0\,\mathrm{erfc}\left\{\left[\left(x^2+y^2\right)/2-\left(x^2+y^2\right)^{1/2}v\,t\right]\left[\frac{4}{3}a_l\left(x^2+y^2\right)^{3/2}\right]^{-1/2}\right\}$$

$$v=Q_p\Big/\left[2\pi\cdot B\cdot N\cdot\left(x^2+y^2\right)^{1/2}\right]+a\cdot v_s$$

式中 Q_p——炼油厂污染水体平均渗入量；

$B\cdot N$——地下水系统岩溶裂隙介质的裂隙率与岩层有效厚度之积；

v_s——地下水天然流速；

a_l——弥散度。

地下水系统石油烃污染物运移预测模型在建立的过程中，受到径向流水质模型的启发，结合某炼油厂地下水系统的实际污染特征与实测资料进行了修正。模型是可靠的，其结果经过实测资料的验证,基本反映了某炼油厂地下水系统石油烃污染质的运移规律，地下水系统石油烃污染预测结果与实测石油烃污染距离的总体误差为3.75%。

3.3.5 地下水石油烃污染物运移预测成果

炼油厂在生产炼制能力一定的条件下,生产污水的排放总量年误差较小,可视为常量。结合野外弥散试验及环境水文地质调查所获得资料，对某炼油厂地下水系统的污染状况进行分析，确定某炼油厂污水的综合渗漏系数为：丰水期 λ_2=0.009 6，枯水期 λ_1=0.008 6。某炼油厂投产后，到1994年原油加工能力变为500 t/a，其后污水渗漏量均保持在1994年基本水平，即污水排放量平均为6 000 m^3/d。以2003年为基准，进行丰水期及枯水期地下水系统石油烃污染物运移预测。预测期分别为2008年和2013年，预测结果如表1所示。

表 1　地下水系统石油烃污染物运移模拟及预测成果一览表

模拟及预测年度	地下水系统石油烃污染物运移方向及污染范围			
	向北(m)	向南(m)	东西向(m)	面积(km^2)
2003 年(枯水期)	1 080	810	920	2.84
2003 年(丰水期)	1 130	875	1 005	3.23
2008 年(枯水期)	1 250	980	1 145	4.08
2008 年(丰水期)	1 310	1 040	1 160	4.56
2013 年(枯水期)	1 450	1 235	1 430	6.37
2013 年(丰水期)	1 660	1 310	1 540	7.17

4　结论

通过某炼油厂地下水系统石油烃污染环境调查与研究，结合地下水动态监测、实验室测试以及有关资料的综合分析，可得以下认识和结论：

(1)某炼油厂在连续 30 多年的生产期间，石油烃类有机污染物通过各种途径渗入地下水系统岩溶裂隙含水层中，在炼油厂区及厂北范围内现已形成了 1.06 km^2 的浮油污染区，且正在不断向西北方向延展。浮油污染的运移规律表现为：随区域地下水系统地下水面形态以及红色黏土充填岩溶裂隙层的层底标高的不同而变化。赋存于地下水系统的浮油已变成了一个稳定的地下水污染源。

(2)应用地质统计方法与数值模拟技术，考虑石油烃类有机污染物在地下水系统含水层中迁移转化的机理建立起的空间二维数学模型，较好地反映了实际的石油烃污染物运移的基本规律。

(3)根据已建立的某炼油厂地下水系统石油烃污染物运移模型预测，某炼油厂地下水系统石油烃类污染域将不断扩大。地下水系统石油烃污染物运移预测结果为：2008 年枯水期地下水系统石油烃污染面积为 4.08 km^2，丰水期地下水系统石油烃污染面积为 4.56 km^2；2013 年枯水期地下水系统石油烃污染面积为 6.37 km^2，丰水期地下水系统石油烃污染面积为 7.17 km^2。

参考文献

[1] Rivett M O, Petts J, Butler B, *et al*.. Remediation of contaminated land and groundwater experience in England and Wales [J].Journal of Water Science and Technology, 2002,65(3):251-268.

[2] 秦煜民,隋智慧,魏有权.应用生物修复技术处理石油污染土壤[J].石油与天然气化工，2002, 31(6):333 - 335.

[3] Lee C H,Lee J Y. Evaluation of air injection and extraction test at a petroleum contaminated site[J].Water, Air and Soil Pollution,2002,135(1):65-97.

[4] Vazquez D,Mansoori G A. Identification and measurement of petroleum recipitates [J].J Petro Sci

Engineer,2000(26):49-55.

[5] Wang Zhendi, Li K, Fingas M, *et al*.. Characterization and source identification of hydrocarbons in water samples using multiple analytical techniques[J].Journal of Chromatography A, 2002,971(1-2):173-184.

[6] Reusser D E, Field J A. Determination of benzylsuccinic acid in gasoline-contaminated groundwater by solid-phase extraction coupled with gas chromatography–mass spectrometry[J]. Journal of Chromatography A, 2002,953(1-2):215-225.

[7] Soga K, Page J W E. A review of NAPL source zone remediation efficiency and the mass flux approach[J]. Journal of Hazardous Materials,2004,110(1-3):13-27.

[8] 窦明，王呈祥，左其亭. 淄博市大武水源地水资源综合评价[J]. 水资源与水工程学报，2005, 16(4): 5-10.

[9] 任增平. 降雨淋滤条件下淄河滩陈化残油释放试验[J]. 环境科学学报, 2002, 22(1): 56-59.

[10] 熊汉辉，易绍金. 极端环境下的石油烃生物降解及其在生物修复中的应用[J]. 油气田环境保护, 2005, 15(3): 21-24.

[11] 任增平，李广贺，张戈. 大武水源地堠皋地区水力截获工程运行的数值模拟[J]. 地下水，2002, 24(1): 14-27.

[12] 朱学愚，刘建立. 山东淄博市大武水源地裂隙岩溶水中污染物运移的数值研究[J].地学前缘，2001, 8(1): 171-178.

[13] 慕山，许端平，陈洪，等. 大庆贴不贴泡周边土壤及地下水石油烃污染规律[J]. 环境科学研究, 2006, 19(2):16-19.

[14] 裴桂红，刘建军，王锦山. 地下水石油类污染物的数学模型及计算方法[J]. 辽宁工程技术大学学报：自然科学版, 2001, 20(2): 240-243.

[15] 刘晓莉，梁冰，薛强. 地下水环境中有机污染物迁移转化动力学模型的研究[J]. 工程勘察，2003, 1: 24-29.

基于ANN与GIS的含水层石油烃污染防污性能研究——以某炼油厂岩溶裂隙水系统为例

方允治，马振民，张亮，李玲玲

(济南大学，济南，250022)

摘　要： 地下水系统石油烃污染因其危害性和复杂性受到普遍关注。本文以某炼油厂所在岩溶裂隙水为研究对象，提出了多因子组合条件下地下水系统石油烃污染防污性能分析的定量化方法，结合选定的评价因子类别确定了 ANN 模型结构，获取各评价因子的权重系数，并运用 ANN 和 GIS 对各因子专题层进行加权重合叠加，提出了地下水系统含水层石油烃污染防污性能分析模型，给出了地下水系统含水层石油烃污染防污性能空间分布结果。

关键词： 人工神经网络；地理信息系统；含水层；石油烃污染；防污性能

随着社会经济的高速发展，石油作为一种重要能源，在国民经济中的地位日益突出。作为社会经济发展的物质基础，石油不但是交通运输、电力、机械制造等行业的重要能源，还是化工、纺织、食品等行业所必须的原料。

石油是一种复杂的多组分混和物，其主要组成是烃类(烷烃、烯烃、芳香烃)，约占总量的 97%～99%，其余是数量不多的非烃组分(含氧化合物、含硫化合物、含氮化合物、胶质和沥青)。石油炼制企业在加工、生产过程中的跑、冒、滴、漏等现象常常会造成地下水系统含水层石油烃含量的迅速升高，水化学类型显著改变。这些严重制约着工农业生产的发展，危害人民的身体健康。因此，对地下水系统含水层石油烃污染防污性能分析及其制图是当前环境科学工作者研究的热点和前沿课题，也是地下水污染防治的基础。而基于 ANN 与 GIS 的技术为解决这一课题提供了一种新的研究方法，其污染性分析成果为自然环境、社会经济等方面提供了管理决策的重要依据。

1　地下水系统含水层石油烃污染特征分析

由以往地下水污染调查和污染规律分析可知，某炼油厂岩溶裂隙水系统含水层石油烃污染有两种潜在污染来源：其一是长期渗漏的炼油厂排放的含油污水；其二是残留在包气带土层中的残油，这两种形式的污染源对地下水的污染过程有所不同。长期渗漏的含油污水，由于下渗量较大，污水可以穿过包气带土层进入含水层而污染地下水，由于对流和弥散作用容易在地下水中形成羽状污染带，而由于石油烃类污染物主要由轻质组分组成，因此主要对上部地下水产生明显的污染。

对于残留在包气带土层的石油烃类污染物，在重力和毛细力的作用下垂直入渗，但由于土层的吸附和滞留作用，其中有相当一部分被截留在包气带土层。在这部分污染物中除含有溶解相石油烃类外，还存在非溶解相石油烃类，这种非溶解相石油烃类因黏滞性大而滞留在污染带上部。当发生降雨时，下渗雨水就会驱赶非溶解相石油烃类，并溶解包气带土层中所吸附的石油烃类进入地下水系统含水层而污染地下水。

2 研究区水文地质条件概况

研究区域位于某省会城市东部，是该市大型水源地补给区之一，区域地下水系统含水层具有一定的防污性能，某炼油厂厂址即坐落于此。研究区在地貌上属于构造剥蚀低山丘陵区和山前倾斜冲洪积平原的过渡带，地形南高北低，三面环山。本区属暖温带大陆性气候，年平均降水量为 650 mm，降水集中在 7 ~ 9 月，占全年降水量的 60%左右。

区域内主要发育有第四系、奥陶系石灰岩及燕山期火成岩三种地层。第四系主要分布在研究区的山间沟谷、山前及北部的广大地区；奥陶系石灰岩在某炼油厂区和以南地区出露地表，厂区以北隐伏于第四系以下。在研究区域内广泛发育的奥陶系石灰岩含水层岩溶裂隙发育，渗透性较强，富水性中等，为研究区域的主要含水层；该石灰岩层倾向北，倾角10°左右。岩溶裂隙水含水层主要接受研究区以南石灰岩裸露补给区域的地下水径流补给及大气降水补给，某炼油厂所处地段地下水流向由南向北，局部为由南西向北东，地下水水力坡度平均值为 2.5‰，地下水排泄以人工开采为主。地下水位埋深 20 ~ 40 m，水位变化受大气降水和人为周期性开采控制，具有枯水期和丰水期周期性交替变化的特征。

3 地下水系统含水层石油烃污染防污性能研究

地下水系统含水层石油烃污染防污性能评价基于人工神经网络(ANN)和地理信息系统(GIS)耦合技术。人工神经网络(ANN)具有大规模并行、分布式存储和处理、自组织、自适应能力，特别适于处理需要同时考虑许多因素和条件的不精确和模糊的信息处理问题；地理信息系统(GIS)具有强大的空间查询、空间分析，以及对数据空间属性的表达和存储的独特功能。将 ANN 与 GIS 耦合可以很好地发挥出其各自的优势。ANN 与 GIS 耦合技术，就是利用 GIS 强大的空间信息处理功能，对各种图形信息进行量化，提取数据，为 ANN 提供必需的输入和训练条件，用 ANN 分析模型对数据进行处理，采用非线性的方式描述各评价因素之间的关系，确定其决策权重，再应用 GIS 的空间查询和分析功能对 ANN 的输出进行处理，以可视化地图的形式给出地下水系统含水层石油烃污染防污性能在空间分布的结果。

3.1 影响石油烃污染防污性能的主控因素

主控因素的恰当选择是对地下水系统防污性能本质的刻画，它不仅决定了预测神经网络模型的输入层结构，同时也影响着分析评价的准确性。事实上，地下水系统含水层石油烃污染防污性能可以理解为含水层对石油烃污染的防污性能反应程度，即抵御石油烃污染的能力。

为了从本质上刻画地下水系统含水层石油烃污染防污性能情况，本文在详细研究分析了影响研究区含水层石油烃污染防污性能的诸多影响因素的基础上，选择了对含水层石油烃污染防污性能有重要影响的包气带介质性质(I_1)、地下水位埋深(I_2)、含水层净补给(I_3)、含水层渗透性(I_4)作为四大主控因素。

3.2 主控因素影响分析及其量化

主控因素影响分析及其量化过程即是对影响地下水系统含水层石油烃污染防污性能的各主控因素进行量化。根据各主控因素对地下水系统含水层石油烃污染防污性能影响程度的大小，对研究区内各主控因素值进行具体量化，利用 GIS 对各种主控因素建立子专题图层，并进行图形信息空间量化处理分析。各主控因素的含义及量化方法如下。

3.2.1 包气带介质的影响

包气带是指地下水位线以上的非饱和区或非连续饱和区。包气带介质的类型决定了渗流路径的长度和路线，以及石油烃污染物在迁移过程的稀释、吸附、降解以及物理化学反应等。包气带内的任何裂隙均对渗流路线起控制作用。研究中，对包气带介质影响强度的量化是根据研究区不同评价单元内包气带介质的性质特征进行划分的，量化时选择对防污性能有显著影响的介质层。当有多层介质时,考虑各层介质的相对厚度以及各层介质对防污性能的影响大小。量化结果如图 1 所示：第一，研究区北部包气带介质中黏土层稳定分布，平均厚度为 5~8 m，且黏土层以下石灰岩风化带中岩石的节理与裂隙多被黏土充填，一般充填带厚度为 10 m 左右，包气带介质中黏土层和充填带较厚且透水性很弱,所以能够作为阻隔石油烃污染物向下运移的良好屏障，有利于提高地下水系统含水层防护能力，其赋值为 1.0；第二，对于包气带介质为固结岩石的,考虑其裂隙、节理、溶洞的发育程度进行量化，由于研究区中南部包气带介质为奥陶系石灰岩，该层石灰岩岩性较脆，节理及岩溶裂隙发育，对地表含油污水防污能力较弱，因而对地下水系统含水层防污性能的影响很大，故将其赋值为 6.0；第三，研究区中部地区包气带介质上层是黄土状亚砂土，下层是红黏土，在这种二元结构中由于红黏土能够有效地限制石油烃污染物向含水层迁移,其对防污性能的影响较大，故选择红黏土作为包气带介质,将其赋值为 2.0；第四，研究区内第四系黄土状亚砂土广泛分布，该层发育较为均匀，上部含沙量较高，向下黏质成分增高。由于黄土状亚砂土限制石油烃污染物向含水层迁移的能力介于红黏土和石灰岩之间，则其赋值为 4.0。

3.2.2 地下水位埋深

地下水位埋深决定着石油烃污染物到达地下水系统含水层之前所经过的距离及与周围介质接触的时间。通常,地下水位埋深越大,石油烃污染物到达含水层所需时间越长,石油烃污染物稀释机会就越多，这同样适用于承压水含水层。显然,潜水含水层由于接近地表,最易被污染。承压水含水层具有天然防止石油烃污染物从地表渗入的性质,其防污性能较强。对于承压水含水层，地下水位埋深指地表到含水层顶板即承压层底板的深度。对于潜水含水层，地下水位埋深则指地表至等水位线的深度。根据地下水位埋深的大小将研究区进行划分，量化结果如图 2 所示：第一，对于地下水位埋深小于 25 m 的地区，其赋值为 5.0；第二，对于地下水位埋深为 25~30 m 的地区，其赋值为 4.0；第三，对于地下水位埋深为 30~35 m 的地区，其赋值为 3.0；第四，对于地下水位埋深大于 35 m 的地区，其赋值为 2.0。

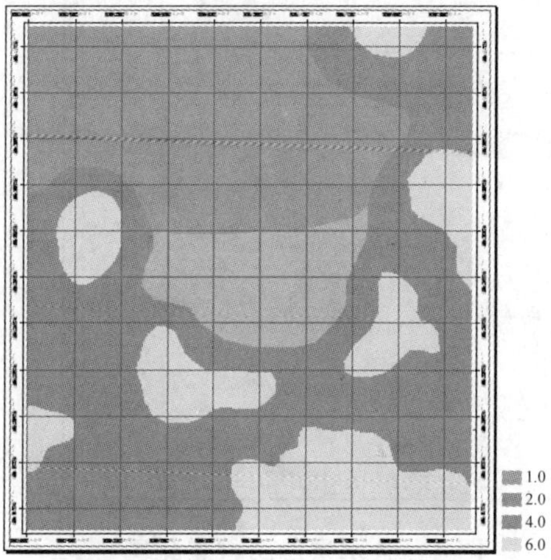

图 1　包气带介质影响量化图

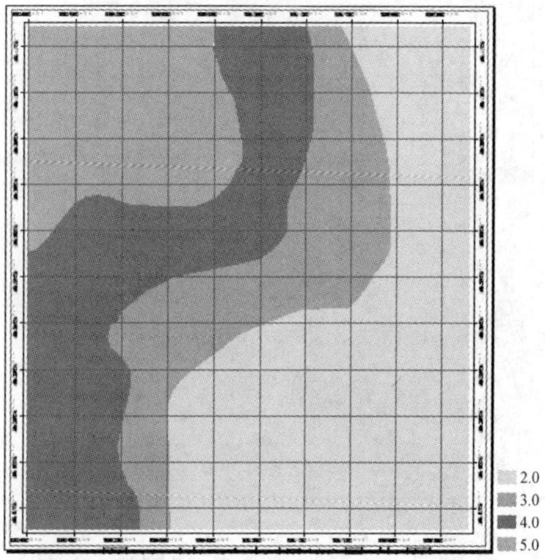

图 2　地下水埋深量化图

3.2.3　含水层净补给

含水层净补给是指单位面积内渗入地表并达到含水层的总水量，一般净补给量指年均补给量。石油烃污染物通过补给水垂直传输到地下水系统含水层并在含水层内水平运移，因此补给水是石油烃污染物运移到含水层的主要载体。补给量越大,地下水污染的潜在趋势越大。研究区属暖温带季风大陆性气候，多年平均降水量为 650 mm。量化过程中对于净补给影响程度的大小依据研究区年均降水量乘以评价单元的有效入渗系数来划分,如图 3 所示：第一，区域内的裸露石灰岩和第四系厚度覆盖较薄的部位内，有效补给量约为 273 mm/a，大气降水可以直接渗入地下水系统，对地下水污染的潜在趋势较大，将其赋值为 6.0；第二，研究区向北第四系厚度逐渐增大，完全被第四系覆盖的北部地区有效补给量约为 143 mm/a，地下水被污染的潜在趋势相对较小，将其赋值为 3.0；第三，位于石灰岩裸露地区和第四系完全覆盖之间的过渡带，有效补给量约为 208 mm/a，地下水被污染的潜在趋势介于二者之间，则将其赋值为 4.0。

3.2.4　含水层渗透性

石油烃污染物在含水层中的运移路线及运移路径的长度由含水层介质控制,其又影响着石油烃污染物的稀释、吸附、分散、降解程度。一般而言,组成含水层介质的颗粒越大或裂隙和溶洞越大,其渗透性越强,含水层的防污性能越弱。研究区内奥陶系石灰岩含水层稳定发育，奥陶系石灰岩是研究区的主要含水层。本文研究中考虑石灰岩含水层中节理裂隙的发育程度不同,根据含水层渗透性的相对差异情况进行量化评分,如图4 所示：第一，对于含水层石灰岩节理及岩溶裂隙发育，渗透性相对较大地区，赋值为 4.0；第二，对于研究区东南部石灰岩含水层，由于火成岩的侵入，在一定程度上阻碍了地下水的运移，起到了相对阻水的作用，所以含水层防污性能较强，将其赋值为 2.0；第三，对于含水层渗透性大小介于上述两种含水层之间的过渡地区，由于含水层中部分节理裂隙被充填，含水层节理及岩溶裂隙一般发育，含水层防污性能亦介于两者之间，故将这些地区赋值为3.0。

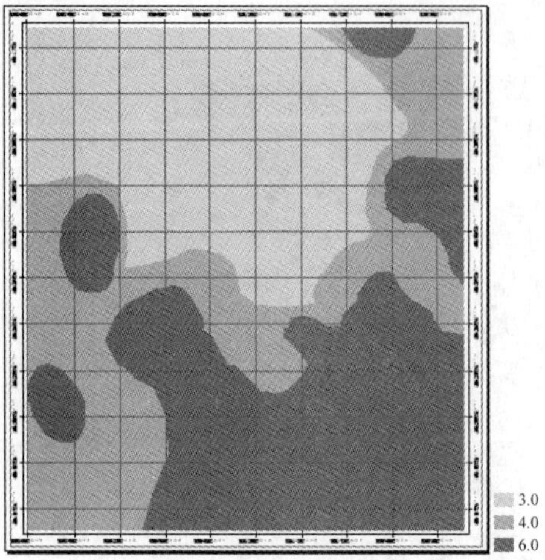

图3　净补给影响量化图

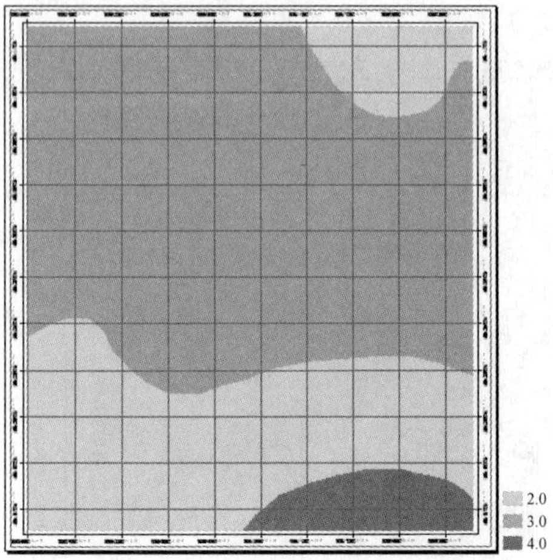

图4　含水层渗透性量化图

3.3　提取训练样本点数据

对 ANN 进行训练时，首先要提供一组训练样本，其中的每个样本由输入样本与理想输出对组成。根据研究区地下水系统含水层防污性能的特征进行专题数据调查，逐一提取各主控因素的量值，在此基础之上对各主控因素的量值分别进行评分，最终形成20套输入输出训练集数据序列($x_1, x_2, \cdots, x_n$；$t_1, t_2, \cdots, t_m$)，为ANN训练提供必需的输入和训练条件。

3.4　通过ANN确定主控因素权重系数

根据对研究区含水层防污性能特征的综合研究，采用改进的有选择的BP网络模型有三层，即输入层、隐含层和输出层。输入层和隐含层各有4个神经元，输出层有一个神经元。其网络结构如图5所示。

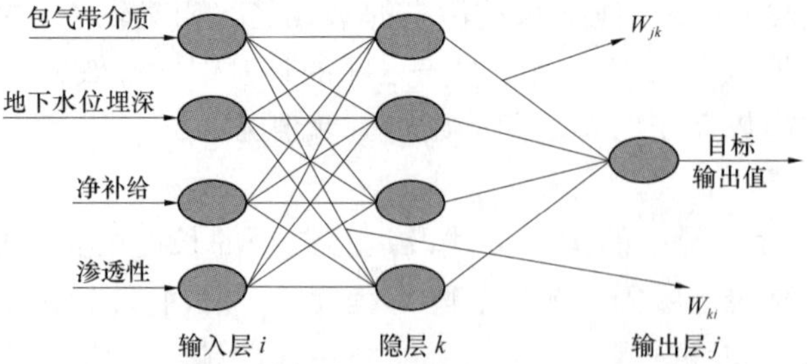

图5　BP人工神经网络结构示意图

评价过程中，取样本点数据对神经网络进行训练。根据训练集数据序列，利用人工神经网络的自适应与学习功能，以简单函数的形式进行多次迭代，实现对映射函数f的逐次逼近，最终得到满意的预测模型。当所有实际输出与其理想输出的误差均方根 $E_{av} \leqslant 0.01$ 时，表明训练结束；否则，通过修正权值，使网络的实际输出与理想输出一致。取隐含层及输出层的学习率均为 0.6，当经若干次训练后，其误差均方根 E_{av}=9.986 0×10^{-3}，达到精度要求，训练结束，各神经元间的连接权值得以确定。

根据下列式(1)~式(3)对各神经元之间的权重加以分析处理，便可得到输入因素对输出因素的决策权重，即权重系数(见表 1)。结果表明，研究区地下水系统含水层防污性能明显受水文地质条件控制。包气带介质、地下水位埋深和含水层渗透性特征是影响地下水系统含水层防污性能的最主要因素，这与研究区内含水层防污性能特征是一致的。

3.4.1 显著相关系数

$$r_{ij} = \sum_{k=1}^{n} W_{jk}\left(1-\mathrm{e}^{-x}\right) \Big/ \left(1+\mathrm{e}^{-x}\right) \tag{1}$$

式中，$x=W_{jk}$。

3.4.2 相关指数

$$R_{ij} = \left|\left(1-\mathrm{e}^{-y}\right) / \left(1+\mathrm{e}^{-y}\right)\right| \tag{2}$$

式中，$y=r_{ij}$

3.4.3 绝对影响系数

$$S_{ij} = r_{ij} / \sum_{i=1}^{m} R_{ij} \tag{3}$$

表 1 评价因子对地下水系统含水层防污性能影响的权重系数

评价因子	包气带介质(I_1)	地下水水位埋深(I_2)	含水层净补给(I_3)	含水层渗透性(I_4)
权重值	0.340 35	0.290 06	0.121 64	0.247 95

3.5 应用 GIS 对多源空间地学信息进行复合叠加

利用 GIS 的强大的空间分析功能，将包含有各主控因素关键属性的同一地区、统一比例尺的多个专题图层进行空间复合叠加，生成一个新的数据层，新数据层的各要素具有各叠加专题层要素的多重属性和各叠加专题层要素属性的统计特征。

3.6 地下水系统含水层石油烃污染防污性能指数计算

引入含水层防污性能指数 S 来对地下水系统含水层石油烃污染防污性能进行评价，防污性能综合评价指数的定义为某单元上的各种主控因素对地下水系统含水层防污性能的影响总和，可用如下模型描述，即

$$S = \sum S_i \cdot I_i$$

式中 S——地下含水层石油烃污染防污性能指数；

I_i——第 i 个主控因素的相对权重；

S_i——第 i 个主控因素的防污性能程度。

在 ANN 与 GIS 耦合模型中，选择包含有各主控因素的规则的、区域一致的多个专题图层的属性数据进行数学计算，将各主控因素专题层的评价单元量化分值分别与各自的权重系数值相乘，再按上述公式得出研究区各评价单元的地下水系统含水层防污性能指数值。

3.7　确定分区阈值

根据上面计算得出的研究区内各评价单元的含水层石油烃污染防污性能指数，利用GIS强大的统计特征，通过GIS的属性统计功能，可以得到地下水石油烃污染防污性能指数的累计频数统计纵向直方图(见图6)，依据直方图的统计分析特征确定地下水系统含水层石油烃污染防污性能的分区阈值。即防污性能弱的分区 $S>7.5$、防污性能较弱的分区 $6.4<S<7.5$、防污性能中等的分区 $5.4<S<6.4$、防污性能较强的分区 $4.3<S<5.4$ 以及防污性能强的分区 $S<4.3$。

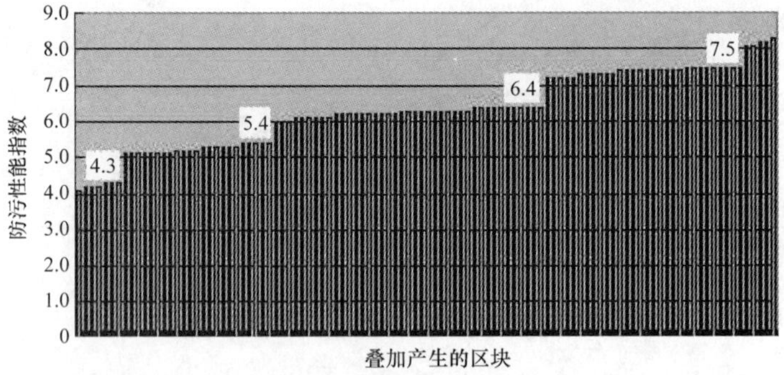

图6　含水层防污性能指数的累计频数统计直方图

3.8　地下水系统含水层石油烃防污性能分区评价

利用GIS强大的空间分析功能，根据上述频数统计的直方图的阈值，定义地下水系统含水层石油烃污染防污性能的分区上下限，将研究区划分为防污性能强、防污性能较强、防污性能中等、防污性能较弱和防污性能弱五个区域，从而得到地下水系统含水层污染防污性能的评价分区成果如图7所示。

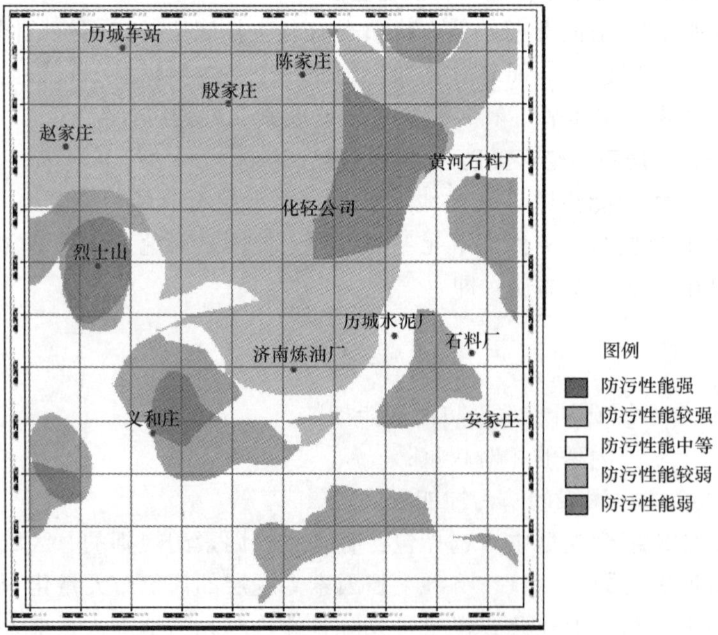

图7　含水层防污性能分区成果示意图

4 结论

ANN 与 GIS 耦合模型在地下水系统含水层防污性能评价中的应用,不仅充分利用了大量复杂的空间信息，而且以非线性方式研究含水层防污性能问题，更加接近问题的本质，使结果更为直观、精确。ANN 与 GIS 耦合技术为地下水系统含水层防污性能分析评价提供了很好的解决方案。评价模型对某炼油厂地下水系统含水层石油烃污染防污性能评价效果理想，评价结果较好地反映了该地区地下水石油烃污染情况，与实际监测资料吻合。因此，将 ANN 和 GIS 技术应用于含水层石油烃污染防污性能评价预测是可行的，具有较高的可信度。

参考文献

[1] 陈葆仁. 人类活动对地下水的影响[J]. 水文地质工程地质，1996(2): 1-4.

[2] 阎平凡，张长水. 人工神经网络与模拟进化计算[M]. 北京：清华大学出版社, 2000: 1-32.

[3] 陈守余，周梅春. 人工神经网络模拟实现与应用[M]. 武汉：中国地质大学出版社, 2002: 1-36.

[4] 付素蓉，王焰新，蔡鹤生.城市地下水污染防污性能分析[J]. 地球科学：中国地质大学学报，2000, 25(5): 477-481.

[5] 向喜琼，黄润秋. 基于 GIS 的人工神经网络模型在地质灾害危险性区划中的应用[J]. 中国地质灾害与防治学报, 2000, 11(3): 23-27.

[6] 陈佩佩，武强. 基于人工神经网络的地裂缝危险性评价系统[J]. 煤田地质与勘探, 2001, 29(3): 44-47.

[7] 曹丽文，姜振泉. 人工神经网络在煤矿开采沉陷预计中的应用研究[J]. 中国矿业大学学报，2002, 31(1): 23-26.

[8] 李新春，孙艳，陶学禹. 应用神经网络评价矿区可持续发展[J]. 中国矿业大学学报，2001, 30(4): 392-395.

城市屋面雨水水质及处理方法探讨

孙小滨，张沛沛，徐玉，刘品

(济南大学，济南，250022)

摘 要： 本文综述了我国南北方城市屋面雨水水质的特点，分析了城市屋面雨水中各污染物浓度特征，并探讨了雨水径流中污染物浓度与降雨量、降雨强度、降雨历时、屋面材料等的关系。在此基础上，综述了国内屋面雨水处理的一些方法及效果，并比较了各方法的利弊。为利用城市屋面雨水回灌地下水提供依据。

关键词： 城市屋面雨水；污染物特性；处理方法

总体上说，我国水资源的分布是南方多于北方，东部地区多于西部地区。由于水资源分布的差异以及我国水资源污染的日益加重，我国许多城市的水资源正在面临着不足和短缺等问题。[1]一方面，由于城区扩张，不透水面积逐年增加，使城市的雨水径流峰值明显增大，流出时间缩短，导致了城市洪涝灾害，对城市的生产和人民生活造成极大的影响，并由此带来巨大的经济损失；另一方面，在我国许多水资源严重缺乏的城市，由于城区人口的激增，需水量也逐渐增大。而低效率的地表水利用率也使城市只能靠超采地下水勉强维持。地下水的连年超采导致地下水位大幅度下降，漏斗面积和深度不断扩大，并带来地下水咸水界面下移、水质污染、海水倒灌等一系列问题。全国目前在北京、河北、上海、天津等省市已形成大范围的地下漏斗区，并因此导致地面沉降、地面塌陷、海水入侵等后果[2-5]。因此，在水资源匮乏的今天，开发利用雨水资源进行地下水回灌已尤为重要。不仅能够缓解水资源匮乏的现状，而且可以减轻暴雨时洪涝灾害的发生。

屋面作为城市不透水面积的一个重要组成部分，其面积在城市面积中占有相当的比重。并且城市屋面雨水便于收集，水质较道路、绿地径流水质好。因此，在水资源日益匮乏的今天，利用屋面雨水回灌地下水具有很大的开发潜力。

1 城市屋面雨水回灌地下水的可行性分析

由于城市环境因素和汇水面材料的影响，城市不同下垫面汇集的雨水受到不同程度的污染。根据汇水面性质的不同，雨水径流的水质差别很大。可以粗略地把城区汇水面分为道路(人行道和广场等)、绿地和屋面三类。

道路汇集的雨水水质较复杂，初期径流污染较严重，COD、SS(固体悬浮物)浓度等主要指标比屋面雨水成倍增高[6]。另外，杨新民认为降雨时交通活动并未停止，随时都有污染物的输入，这使得路面径流的 TP、SS 有时却随降雨的进行而增加[7]。若利用路面雨水进行回灌，则在回灌前须进行严格的处理措施，而这势必会增加工程运营的成本。

因此，从经济角度来看，利用路面雨水进行回灌地下水是不可取的。

由于植物和土壤的吸收拦截，绿地汇集的雨水径流较好，但在绿地或半裸露土地的植被覆盖差的情况下，汇集后的雨水很可能携带大量的泥沙，水质较差，从而对回灌设施造成不同程度的堵塞；另外，绿地雨水收集效率(径流系数)低，因此从汇集量上来说也是不可行的。

一般认为屋面雨水属于轻污染型雨水，其水质较好、径流量大、便于收集利用，其利用价值最高，有利于回灌。回灌前要了解当地城市屋面雨水径流的污染物特征及影响因素。

2 城市屋面径流的污染特征

相比较而言，城市屋面径流的水质好于绿地与路面径流的水质，但就我国各城市屋面雨水水质监测结果来看。南、北方城市屋面雨水的水质却有很大的不同，北京、上海屋面径流污染物中的 COD、SS 的监测结果[8]如表 1 所示。

表 1 北京、上海城市屋面雨水径流 COD、SS 浓度

污染指标	上海		北京	
	沥青屋面径流	瓦屋面径流	沥青屋面径流	瓦屋面径流
COD (mg/L)	30	59	136	136
SS (mg/L)	72	8	328	123

从表 1 可以看出，无论是哪种屋面材料，上海市屋面径流中的 COD、SS 的值都比北京的值低很多，说明上海市屋面径流中的 COD、SS 好于北京市屋面径流的。原因是上海市降雨充沛，大气质量好于北京。在一定程度上也反映出南方雨水径流水质要好于北方。

3 影响城市屋面雨水水质的因素

影响径流水质的因素有很多，且比较复杂，综合起来主要受屋面材料、降雨量、降雨历时、降雨强度、两次降雨间隔等因素的影响[9]。另外，城市的空气质量也是不可忽略的因素。而且雨水径流水质还受到气温的影响，其表现特征是屋面雨水水质呈现季节性变化规律。

3.1 屋面材料对雨水径流水质的影响

建筑屋面材料有水泥保护层屋面、砾石保护层屋面、瓦屋面、石油沥青油毡面等，我国大多数建筑物都采用沥青油毡(平顶)和瓦屋面(坡顶)，研究表明，油毡屋面径流明显比瓦屋面污染严重[10]。

张戈、明雨研究大连屋面雨水水质规律时发现[11]，瓦屋面与沥青油毡屋面径流水质的差异如图 1 所示。从图中可看出两种不同材料屋面初期径流的 COD 浓度相差 2~5 倍，随着降雨历时的增加，COD 可降到 100 mg/L 以下。纪桂霞等在研究上海市屋面雨水水质时(见图3) [12]，发现各场次降雨瓦屋面雨水污染物浓度比沥青屋面低。从而可以看出屋面材料的不同对水质的重要影响。

3.2 屋面雨水径流与降雨量及降雨强度的变化关系

降雨量和降雨强度是影响屋面径流水质的重要因素。它们对屋面污染物具有冲刷、稀释和溶解等多重作用，一般降雨量及降雨强度大，冲刷力度大，初期径流中污染物浓

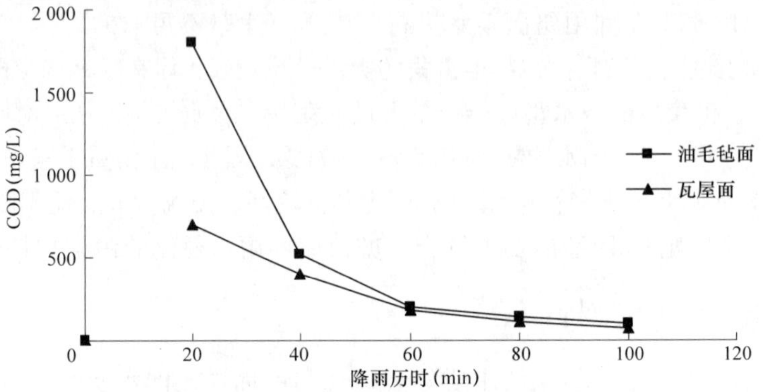

图 1　不同材料屋面径流水质比较

注：引自张戈等对大连屋面雨水水质规律的研究。

度就会高，但很快能够达到稳定[13]。实际上，由于降雨强度的不同，径流污染物中的悬浮物颗粒大小也不同，只有当降雨强度大时，大颗粒的悬浮物颗粒才能被冲下来。因而，笼统地说降雨强度越大污染物浓度越低是不准确的。

3.3　屋面雨水径流与降雨历时的变化关系

一般情况下，在降雨产流的过程中，初期径流中污染物的浓度最高，污染物浓度变化趋势为随降雨历时的增加而减小，这是由于雨水对屋面的冲刷作用。一般说来，相邻两场降雨的间隔时间越长，则累积的污染物量就越大，降雨发生时，初始浓度势必会很高，随着降雨历时的增加，污染物浓度呈下降趋势，最后趋于稳定。上海市屋面雨水水质监测结果见图 2，得出屋面雨水径流水质的 COD、NH_3–N、SS、DS(溶解质)随降雨历时增长逐渐下降，并最终趋于稳定的结果。

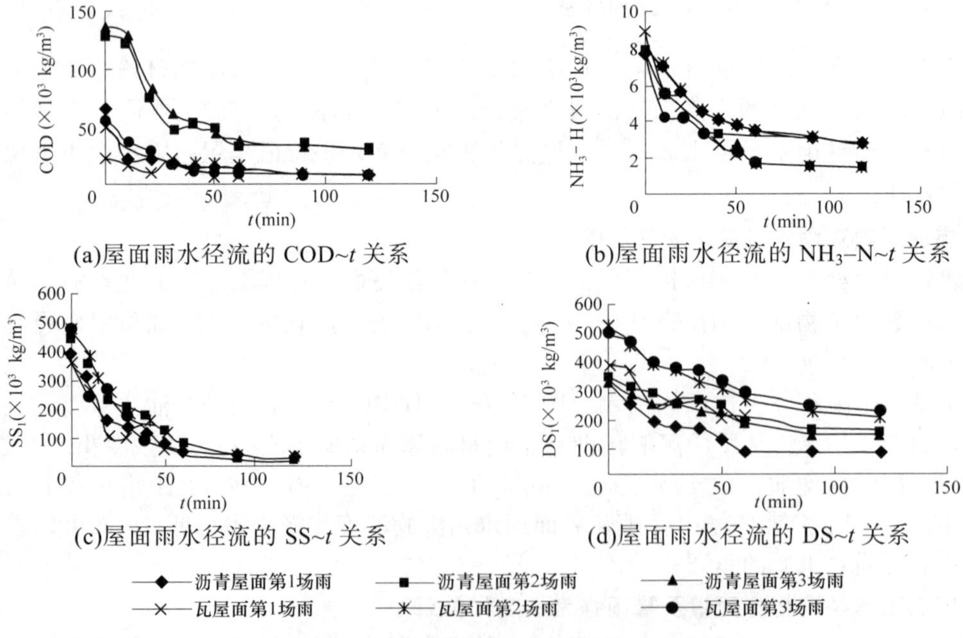

图 2　上海市屋面雨水水质监测结果

注：引自纪桂霞等对上海市屋面雨水水质的研究。

3.4 屋面雨水径流水质与气温的关系

研究表明气温对屋面径流水质也有重要的影响，这种现象在油毡型屋面尤为明显。车武等通过大量的雨水水质分析表明，屋面径流水质有明显的季节性。一般来说，屋面最严重的污染发生在每年春季最初的几次降雨和夏季高温期。[13]前者污染物浓度高的原因是由于几个月没有降雨，使得屋面积累了很多污染物，因此在春季的前几次降雨中，其污染物浓度会较高。夏季高温时，在强烈的日光照射下，沥青油毡析出相当量的有机物溶入径流中，其成分较为复杂。一般日照越强烈，气温越高，沥青油毡屋面材料的分解就越明显，相应屋面雨水径流的 COD 越高，主要为溶解性的难降解有机物，BOD_5/COD 值一般为 0.1~0.2，因此可生化性差。总之，夏季降雨量多，污染物浓度最低；由于冬季和秋季降水少，其浓度次之；春季屋面雨水污染物浓度最高。

4 常规的雨水处理方法

由于城市环境因素和汇水面材料的影响，城市屋面雨水受到一定程度的污染，属于轻污染型，雨水中含有 SS、硝酸盐、亚硝酸盐、微量重金属等污染物质，尤其在初期雨水中浓度较高时，如果将这些雨水直接回灌地下势必污染地下水源，因此若利用屋面雨水进行回灌时首先要进行初期弃流，可在雨水管底部安装初期弃流装置。一般来说，初期弃流量约为 2 mm[14]，作者认为弃流量应根据实际情况而定，主要取决于当地的屋面材料和空气质量；在选择弃流量时应从技术和经济的角度来衡量，若弃流量少，则前期预处理时投入的资金可能会较多，而且技术上需要更严格的处理工艺。若弃流量大的话，会造成水资源的浪费，也是不可取的。对于济南市，从年降雨量及大气状况出发，作者认为弃流量可选 4~5 mm。弃流后的雨水在回灌前需进行处理，各污染物浓度达到地下水回灌标准或饮用水标准后再进行回灌。因此，利用雨水人工补给地下水技术中最重要的问题是雨水净化。

一般来说，处理屋面雨水的方法可分为物化处理、生化处理、生态处理和膜过滤。车武等研究发现表明屋面径流可生化性差，因此不宜采用生化处理。从技术上来讲，可采用物化处理法、生态处理法或膜过滤法。

4.1 物化处理法

物理化学处理法(物化法)是废水处理方法之一，是运用物理和化学的综合作用使污水得到净化的方法。包括物理过程和化学过程的单项处理方法，如浮选、吹脱、结晶、吸附、萃取、电解、电渗析、离子交换、反渗透等[15]。无论是从技术角度还是从经济角度来看，物化处理都是很好的屋面雨水处理方法。

为去除悬浮的和溶解的污染物而采用的化学混凝–沉淀和活性炭吸附的两级处理是一种比较典型的物理化学处理系统。此法占地面积少；出水水质好，且比较稳定；对污染水水量、水温和浓度变化适应性强；可去除有害的重金属离子；除磷、脱氮、脱色效果好；管理操作易于自动检测和自动控制等。图 3 显示了屋面雨水径流物化处理流程图。根据滤料的不同，可设计不同的过滤系统处理屋面雨水。

何江涛、陈鸿汉等提出的人工快速渗滤系统(Constructed Rapid Infiltration)[16]，简称 CRI，是一种先进的污水土地处理技术，该系统的渗滤池为人工填充的天然介质，主要

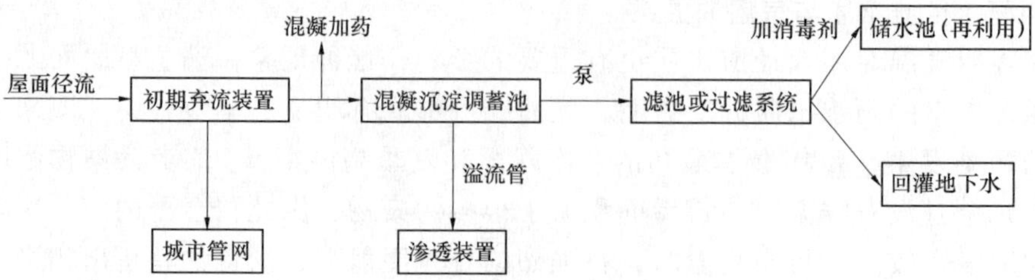

图 3　屋面雨水径流物化处理流程图

由河流冲积砂颗粒组成，其砂粒直径小，比表面积大，介质孔隙度小，并掺入一定量的特殊填料，通过渗滤介质和微生物的物理、化学、生物吸附、过滤与净化作用和自我调控功能，使污水得以净化。经 CRI 处理后，雨水中的 SS 等颗粒型污染物可通过快渗滤池过滤来去除，其中 COD_{Cr}、BOD_5 的去除率分别达到了 70%~90%、70%~80%，TN、NH^+_4 的去除率分别为 70%~95%、80%~95%，可溶性有机污染物则通过好氧微生物和厌氧微生物降解，并能达到中水回用的标准。在北京石景山京原中学进行野外试验时，达到了预期的效果，处理的雨水用于冲洗操场、学校绿化及冲厕等。但处理后的 COD_{Cr}、TN 不稳定，有时无法达到回灌地下水水质标准。该方法的优点是经处理后出水水质较好，且运行稳定，投资少，管理方便；缺点是占地指标稍差，且在实际中可人为调控的因素少，难以进行全面的研究，且实验室与野外试验结果的可比性差，难以确认各因素的影响机理。因此，若将处理后的雨水回用，则是一种易于推广的雨水处理措施，但如果进行地下水回灌，由于处理后的水质无法达到回灌地下水的要求，因此需要进一步改善处理工艺，使之达到国家地下水水质标准或饮用水标准。

田景宏等提出的利用砂滤人工湿地系统处理污染河水[17]，取得了一定的处理效果，创造了良好的生态景观效果。而范平、吴纯德提出的 GAC–石英砂生物滤池处理微污染原水也取得了不错的效果[18]，该方法利用原有的砂滤池在石英砂上添加活性炭滤料来实现，此法可利用石英砂的物理拦截作用和活性炭的吸附作用去除雨水中的胶体颗粒物和有机物，有效改善水质。该方法的优点是：利用该工艺处理后的雨水水质中的 NH_3–N、COD_{Mn}、浊度平均值能达到 0.1 mg/L、1.56 mg/L、0.23 NTU，出水中的亚硝酸盐氮含量几乎监测不出来，能达到回灌水水质的要求。缺点是：若进水水质较差，则预处理出水中含有丰富的生物群落，若气候炎热，在高温高湿季节，会经常出现藻类繁殖，为红虫的繁殖提供了有利的条件，水池池底易生成淤泥，滋生苔藓，需经常进行清淤，而且活性炭价格较昂贵，投资较大。

作者认为仅仅利用砂滤处理屋面雨水后的水质可能达不到地下水水质标准或饮用水标准。在实际中可利用砂滤+活性炭+砂滤的方法处理屋面雨水。即在炭砂滤池的活性炭上层可再加一层砂层，将大部分悬浮物、浊度等污染物去除，然后利用活性炭的吸附作用去除砂滤无法去除的污染物，如挥发酚、氨氮、重金属等。通过试验可确定活性炭及上层砂层的厚度。上述方法既延长了活性炭的使用寿命，又在实际中可与上层的砂层一起进行定期更换，经济上也可行，是一种值得继续研究的处理工艺。

4.2 生态处理法

雨水生态利用技术是 20 世纪 90 年代开始在德国兴起的一种综合性雨水利用技术。这种方法是将城市径流雨水(路面和屋面)引入到绿地、花坛、池塘、湖泊、湿地、河道等设施来接纳并净化雨水。常用的设施有各种透水性地面、渗透沟管、渗透井、渗透池塘等[19]，以截留、下渗雨水补充城区浅层地下水；利用人工或天然湿地、河道、池塘、湖泊净化雨水来减轻水体的污染。该方法的优点是将直接利用与间接利用相结合，将人工处理与自然净化结合起来，有效地改善了城市水环境与生态系统。并间接补充了浅层地下水。处理效果较好且处理成本低，缺点是占地面积大且难以补充深层含水层。在城市地区不易推广。

4.3 膜过滤法

膜过滤法指以压力为推动力，依靠膜的选择性，将液体中的组分进行分离的方法，包括微滤(MF)、超滤(UF)、纳滤(NF)和反渗透(RO)四种过程。[20]一般说来，膜技术占地小，经过膜直接处理后的水质很好；但工艺设备投资大、维护复杂、处理成本高。因此，在实际地下水回灌工程中难于推广。

5 结语

随着城区面积的扩张，不透水面积的增大，屋面作为城市面积的主要组成部分。屋面雨水势必成为回灌地下水的一种重要水源，但由于屋面雨水的可生化性差，在实际中不宜采用生化处理法，而膜过滤法又经济投入太大，也不是一种合适的方法，因此应加强雨水的处理工艺的研究技术，采用物化法或物化法与膜过滤法，即多种处理工艺相结合的综合方法进行处理。总之,利用屋面雨水进行回灌，涉及到回灌方式、处理技术的实用性、经济效益和管理办法等方面。因此，在实际中要综合考虑到上述因素，避免科研和生产的脱离，对于地下水回灌工程，开发适合当地的屋面雨水处理工艺，更高效地利用雨水资源。

参考文献

[1] 国务院.中国 21 世纪议程——中国 21 世纪人口、环境与发展白皮书[M].北京:中国环境科学出版社,1994.

[2] 张艳红,胡海军,等.河北省雨水集蓄利用工程发展初探[J].南水北调与水利科技,2002,23(5):37-38.

[3] 吴林奎. 上海——世界上第一座下沉的大城市[J].上海党史与党建,1994,4: 23-28.

[4] 孙承志,高云安,等.北京市东郊地面沉降分析[J].岩土工程界, 2002, 11: 31-33.

[5] 王若柏.城市地面沉降的灾害链特征——以天津市为例[J].气象与减灾研究,2008,31(1):54-60.

[6] 车武,李俊奇,等.对城市雨水地下回灌的分析[J].城市环境与城市生态,2001,14(4):28-30.

[7] 杨新民,沈冰,等.降雨径流污染及其控制评述[J].水土保持学报,1997,3:59-63.

[8] 车武,刘燕,等.国内外城市雨水水质及污染控制[J].给水排水,2003,29(10):42-46.

[9] 殷捷.城市屋面径流污染特性及其利用探讨[J].安徽农业科学,2007,35(17):5244-5245.

[10] 曹秀芹,车武,等.城市屋面雨水收集利用系统方案设计分析[J].给水排水,2002,28(1):13-15.
[11] 张戈,明雨.大连市屋面雨水利用中的水质研究[J].化学工程与装备,2008(7):98-100.
[12] 纪桂霞,王平香,等.上海市屋面雨水水质监测与处理利用方法研究[J].上海理工大学学报,2006,28 (6):594-598.
[13] 车武,欧岚,等.北京城区雨水径流水质及其主要影响因素[J].环境污染治理技术与设备,2002,3(1):33-37.
[14] 车武,欧岚等.屋面雨水土壤层渗透净化研究[J].给水排水,2001,27(9):38-41.
[15] 上海市环境保护局.废水物化处理[M].上海:同济大学出版社,2007.
[16] 张国臣,陈鸿汉,等.人工快速渗滤系统在城市雨水资源化中的应用[J].工程技术,2008,390:76-78.
[17] 田景宏,黄柄彬.砂滤人工湿地处理污染河水运行研究[J].水处理技术,2008,34(8):49-51.
[18] 范平,吴春德,等.GAC–石英砂生物滤池处理微污染原生[J].水处理技术,2008, 34 (8):59-62.
[19] 王小君.景观水生态处理技术应用探讨[J].北京园林,2007,23(1):43-46.
[20] 黄明珠,曹国栋,等.膜产品及膜过滤方式介绍[J].中国水利,2006,3:65.

1 Introduction to Managed Aquifer Recharge

1.1 What is managed aquifer recharge?

Managed aquifer recharge (MAR) is the purposeful recharge of water to aquifers for subsequent recovery or environmental benefit. Aquifers, permeable geological strata that contain water, are replenished naturally through rain soaking through soil and rock to the aquifer below or by infiltration from streams. The human activity which enhances aquifer recharge can be put into three categories:

(1)Unintentional—such as through clearing deep-rooted vegetation, by deep seepage under irrigation areas, and in cities also by leaks from water pipes and sewers.

(2)Unmanaged—including stormwater drainage wells and sumps, and septic tank leach fields, usually for disposal of unwanted water without thought of reuse.

(3)Managed—which includes injection wells, and infiltration basins and galleries for rainwater, stormwater, reclaimed water, mains water and water from other aquifers that is subsequently recovered for all types of uses.

This document focuses only on this final category, but acknowledges the opportunities to convert from unmanaged recharge to managed recharge with the aim of protecting the environment and using the recovered water.

Enhancing natural rates of groundwater recharge via MAR provides an important potential source of water for urban and rural Australia. This paper addresses all forms of MAR, but the emphasis is on urban applications.

MAR can be used to store water from various sources, such as stormwater, reclaimed water, mains water, desalinated seawater, rainwater or even groundwater from other aquifers. With appropriate pretreatment before recharge and sometimes post-treatment on recovery of the water, it may be used for drinking water supplies, industrial water, irrigation, toilet flushing, and sustaining ecosystems.

Common reasons for conducting MAR include:

- securing and enhancing water supplies;
- improving groundwater quality;
- preventing salt water from intruding into coastal aquifers;
- reducing evaporation of stored water, or
- maintaining environmental flows and groundwater-dependent ecosystems, which improve local amenity, land value and biodiversity.

Consequential benefits may also include:

- improving coastal water quality by reducing urban discharges,
- mitigating floods and flood damage, or
- facilitating urban landscape improvements that increase land value.

MAR can play a role in increasing storage capacity to help city water supplies cope with the runoff variability in Australian catchments exacerbated by climate change and assist in harvesting abundant water in urban areas that is currently unused.

1.2 Types of managed aquifer recharge

A wide range of methods are in use for recharging water to meet a variety of local conditions. For examples see Figure 1.1 and Figure 1.2.

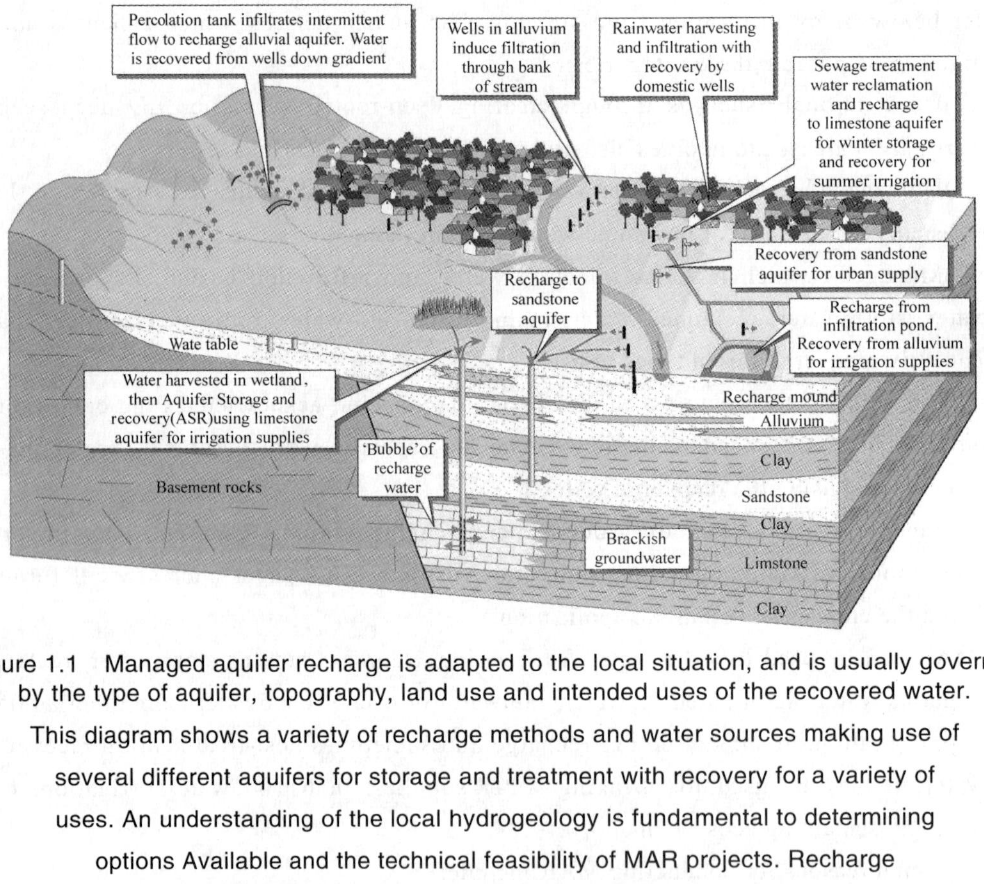

Figure 1.1　Managed aquifer recharge is adapted to the local situation, and is usually governed by the type of aquifer, topography, land use and intended uses of the recovered water. This diagram shows a variety of recharge methods and water sources making use of several different aquifers for storage and treatment with recovery for a variety of uses. An understanding of the local hydrogeology is fundamental to determining options Available and the technical feasibility of MAR projects. Recharge shown here occurs via wells, percolation tanks and infiltration basins (Adapted from Gale, 2005,with permission)

(1)Aquifer storage and recovery (ASR): injection of water into a well for storage and recovery from the same well. This is useful in brackish aquifers, where storage is the primary goal and water treatment is a smaller consideration (for example Paddocks, South Australia).

(2)Aquifer storage, transfer and recovery (ASTR): involves injecting water into a well for storage, and recovery from a different well. This is used to achieve additional water treatment in the aquifer by extending residence time in the aquifer beyond that of a single well (for example Parafield Gardens, South Australia).

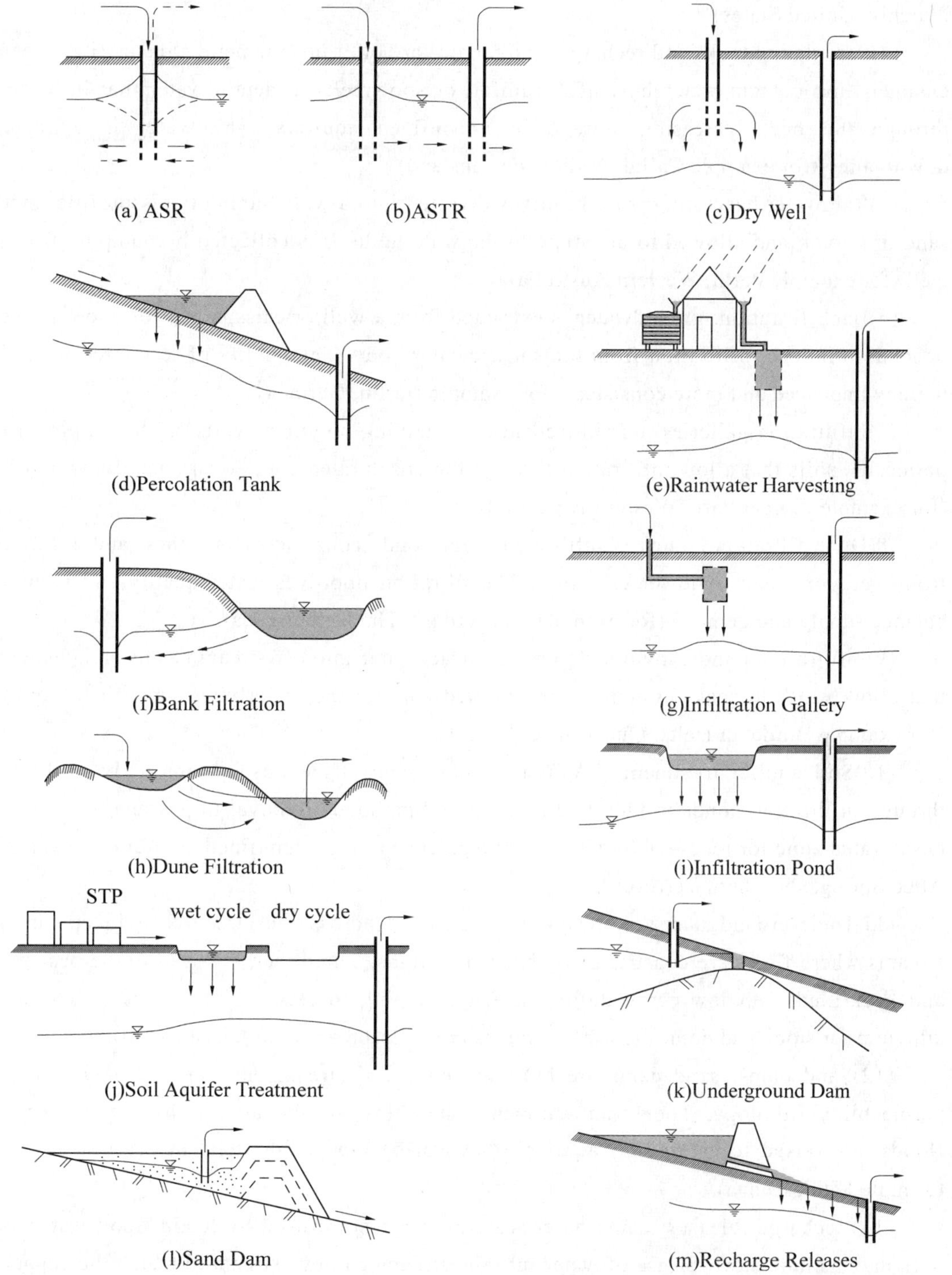

Figure 1.2 Schematic of types of Managed Aquifer Recharge.

(Source: Dillon, 2005, extended in EPHC-NHMRC-NRMMC, 2008b)

(3)Vadose zone or ‘dry’ wells: typically shallow wells in areas with deep watertables. They allow infiltration of high-quality water to the unconfined aquifer at depth (for example Phoenix, United States).

(4)Percolation tanks and recharge weirs: dams are built in ephemeral streams (i.e. stream channels that contain water only after rainfall or snowmelt) to detain water that infiltrates through the bed, increasing storage in unconfined aquifers. The water is extracted down-valley (for example Callide Valley, Queensland).

(5)Rainwater harvesting: roof runoff is diverted into a well, sump or caisson filled with sand or gravel, and allowed to percolate to the watertable. It is collected by pumping from a well (for example Perth, Western Australia).

(6)Bank filtration: groundwater is extracted from a well or caisson near or under a river or lake, to induce infiltration from the surfacewater body. The quality of recovered water is thereby improved and more consistent (for example Berlin, Germany).

(7)Infiltration galleries: buried trenches (containing polythene cells or slotted pipes) in permeable soils that allow infiltration through the unsaturated zone to an unconfined aquifer (for example Floreat Park, Western Australia).

(8)Dune filtration: water is infiltrated from ponds constructed in dunes, and extracted from wells or ponds at lower elevation. The filtration improves water quality and helps to balance supply and demand (for example Amsterdam, The Netherlands).

(9)Infiltration ponds: involve diverting surfacewater into off-stream basins and channels that allow water to soak through an unsaturated zone to the underlying unconfined aquifer (for example Burdekin Delta, Queensland).

(10)Soil aquifer treatment (SAT): treated sewage effluent is intermittently infiltrated through infiltration ponds to facilitate nutrient and pathogen removal in passage through the unsaturated zone for recovery by wells after residence in the unconfined aquifer (for example Alice Springs, Northern Territory).

(11)Underground dams: a trench is constructed across the stream bed in ephemeral streams where flows are constricted by basement highs. The trench is keyed to the basement and backfilled with low-permeability material, helping to retain flood flows in saturated alluvium for stock and domestic use (examples are widespread in northeast Brazil).

(12)Sand dams: sand dams are built in ephemeral stream beds in arid areas on low permeability lithology. They trap sediment when flow occurs and, following successive floods, are raised to create an ‘aquifer’ that can be tapped by wells in dry seasons (for example Kitui, Kenya).

(13)Recharge releases: dams on ephemeral streams are used to detain flood water and uses may include slow release of water into the streambed downstream to match the capacity for infiltration into underlying aquifers, thereby significantly enhancing recharge (for example Little Para River, South Australia).

Selection of suitable sites for MAR and choice of method will depend on the hydrogeology, topography, hydrology and land use of the area. It is common to find similar types of MAR projects clustered in the same area due to shared physical attributes. In another area, the methods may be quite different.

MAR is in wide use in many countries to enhance water supplies, particularly those in semi-arid and arid areas, but also in humid areas, primarily for water quality improvement. The International Association of Hydrogeologists has a Commission on Managed Aquifer Recharge whose web site contains many case studies found in downloadable documents (www.iah.org/recharge/).

1.3 Components of a MAR project

Figure 1.3 shows the seven elements common to all types of MAR projects. However, MAR projects may appear quite different because of the type of aquifer available for storage. There are two main types of aquifers: those that are confined by a low permeability layer, which for MAR requires injecting water via a well (see Figure 1.3(a)), and those that are unconfined and allow water to infiltrate through permeable soils, where recharge can be enhanced by basins and galleries (see Figure 1.3(b)).

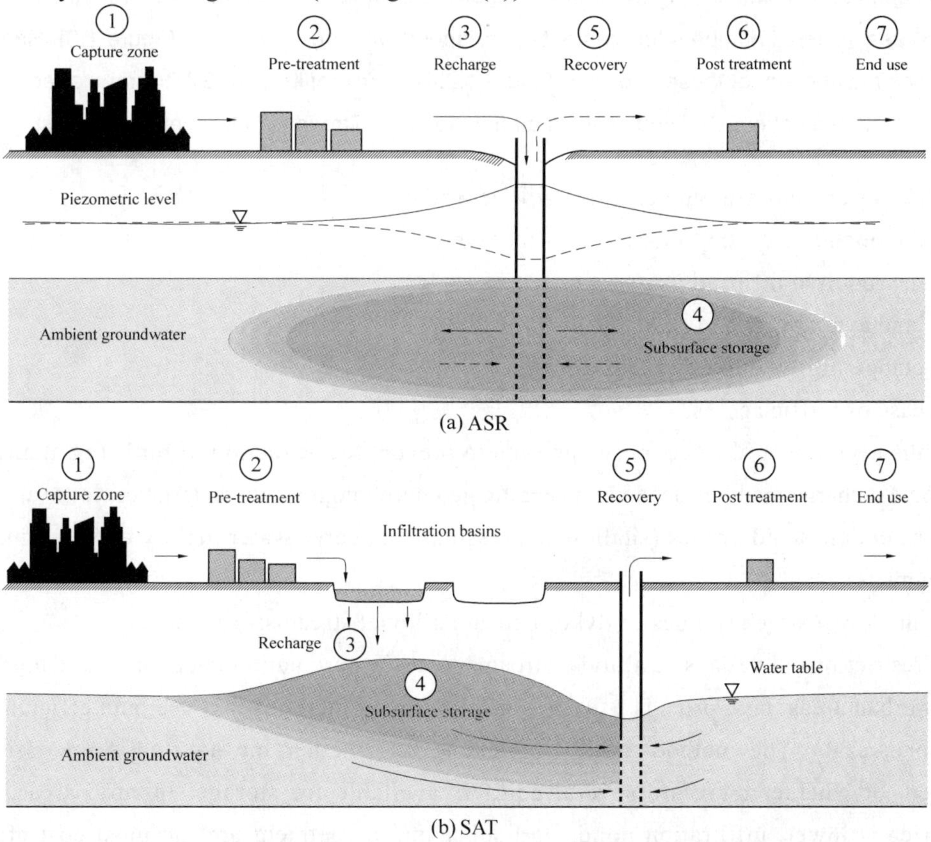

Figure 1.3 Two examples of managed aquifer recharge, (a) ASR and (b) SAT, showing the seven elements common to each system. Piezometric level is the level of water in a well if a well were constructed. For an unconfined aquifer this is the watertable. During recharge levels rise and near recovery wells levels fall. Numbers used in the figures correspond to those shown in Figure 1.4.

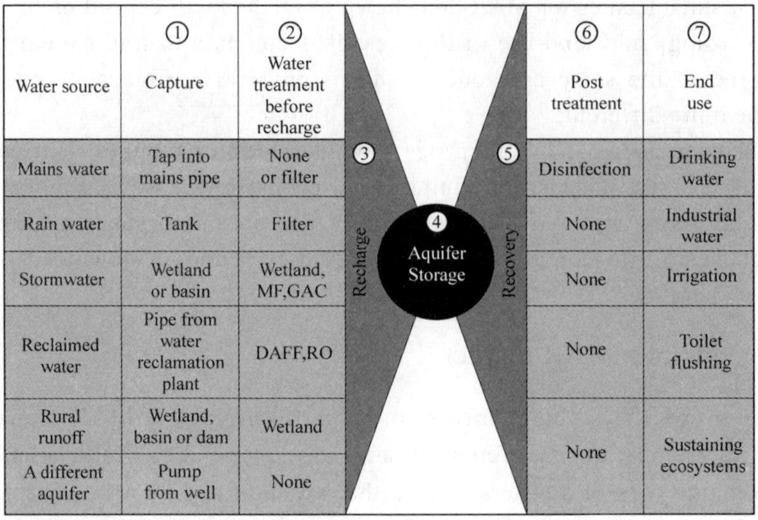

Figure 1.4　Typical sources of water, methods of capture and pretreatment for MAR. All sources of water, in combination with the right treatment before recharge, can be recovered from the aquifer for any end use. (Circled numbers for elements correspond with those shown in Figure 1.3). Recharge methods may be any of those shown in Figure 1.2. MF= microfiltration, GAC=granular activated carbon filtration, DAFF=dissolved air flotation and filtration, RO=reverse osmosis.

The chosen configuration and size will depend on:

- the thickness of the low-permeability layer;
- the required infiltration rate;
- land availability and cost;
- compatibility with other land uses;
- ease of traffic access;
- the need to avoid insect pests, or even to prevent the attraction of birds (eg at airports).

Some recharge methods depend on specific geomorphological characteristics, for example:

- sand dunes and swales (shallow depressions that carry water mainly during rainstorms or snowmelts);
- shallow bedrock saddles or dykes beneath alluvial streams;
- restrictions in valleys of alluvial streams to allow percolation tanks or sand dams.

In urban areas, land price is usually high, favouring methods that use land efficiently (eg ASR, or ASTR). The method used will depend on whether the aquifer is unconfined or confined, or whether there are several aquifers available for storage. In rural areas, where land price is lower, infiltration ponds and soil aquifer treatment are the most cost-effective ways of recharging large volumes of water.

Source-water quality may also play a role in method selection. In general, if the turbidity or nutrient concentration of the source water is high or variable, well injection methods are likely to lead to rapid clogging. In this situation, infiltration basins that can be periodically scraped or ploughed are preferred. Alternatively, if higher treatment levels are required, provision must be made for back-flushing wells to purge them of sediment and biomass, or for discharge or recycling of water-treatment by products.

All these factors must be considered before selecting a recharge method. Investigation costs may be substantially greater for projects involving unconfined aquifers than for confined systems because, in addition to characterising the aquifer, an understanding of the soil profile through to the watertable is needed. In some locations, it may be desirable to store waters destined for different end uses in separate aquifers, and possibly use different methods of recharge.

There are many combinations of water sources, water treatments and end uses (see Figure 1.4). Generally poorer quality source waters will need a higher level of treatment before recharge in cases where:

- the aquifer already contains high quality water,
- the water is to be recovered for higher valued uses such as drinking, or
- the aquifer is fine-grained and there is a need to avoid frequent or permanent clogging of the recharge basin, gallery or well.

For urban stormwater recharging a brackish limestone aquifer, a passive treatment such as a wetland may be suitable for recovery of water for irrigation without any requirement for post-treatment. If recovering for drinking supplies residence time in the aquifer may also be needed to assure quality, with post-chlorination if recovering to mains without diluting the chlorine residual. At a site with a very fine-grained aquifer it has been found that microfiltration (MF) and granular activated carbon (GAC) filtration were needed to prevent clogging of the ASR well and these requirements were more stringent than those to protect groundwater quality and for recovered water to be fit for use. Where river water is diverted to infiltration basins for enhancing irrigation supplies, if turbidity is low no treatment may be necessary. Where reclaimed water is used for recharge to recover for potable supplies, the water will be highly treated prior to recharge, and the new risk-based MAR guidelines for protecting health and environment are briefly summarised later.

1.4 Australian experience in MAR

The diversity and geographic spread of MAR in Australia has increased in recent years and in 2008 five states and territories have operational MAR projects and two states have investigations underway (see Figure 1.5).

MAR began in rural areas of Australia before migrating to urban centres. It is expected that after water resources management restores environmental flows in currently over-allocated catchments, that MAR will again grow in rural areas.

Infiltration basins established in the mid 1960s on the Burdekin Delta, Queensland, are the longest serving and currently the largest (up to 45 GL/a) managed aquifer recharge operations in Australia (Volker (ed) 1981； Charlesworth *et al.*, 2002).

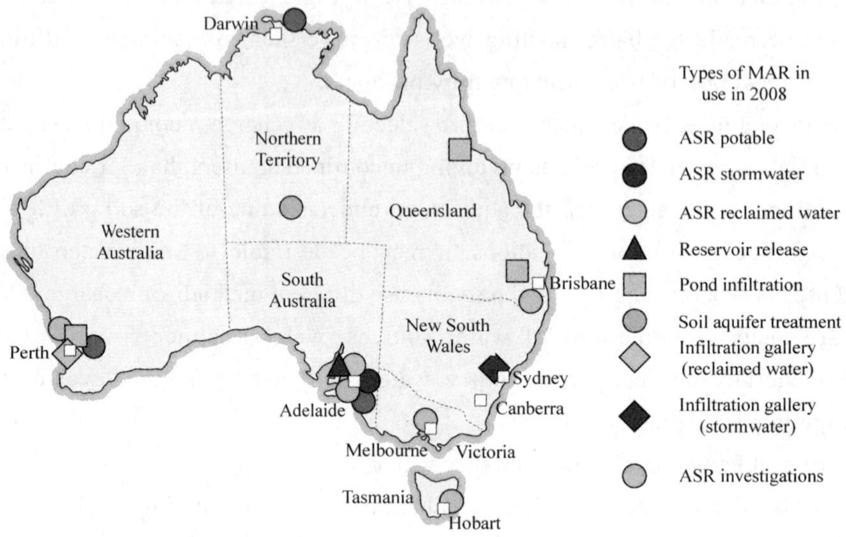

Figure 1.5　Locations and types of MAR in Australia in 2008

Recharge weirs were built on Callide and Lockyer Creeks in the 1970s, and recharge via wells commenced in 1970 in the Angas-Bremer irrigation area of South Australia (SA) and expanded to 30 wells recharging 2.4 GL/a in 1992 (Gerges *et al.*, 2002). Recharge releases (1.5 GL/a) commenced on the commissioning of the Little Para Dam, SA, in 1979 (Dillon, 1984) to substitute for the reduction in natural recharge of Northern Adelaide Plains aquifers from that stream.

In 1992, urban stormwater ASR was initiated at Andrews Farm SA in limestone and in 1994 at Regent Gardens in fractured rock (Gerges *et al.,* 2002). Reclaimed water ASR began at Bolivar SA in 1999, and via infiltration ponds at Halls Head Western Australia (WA) in 2000 (Toze *et al.*, 2002) and via infiltration galleries at Floreat Park WA in 2005 (Bekele *et al.*, 2006). ASR with mains water began at Jandakot WA in 2000 (Martin *et al.*, 2002) and with water from a shallower aquifer at Warruwi Northern Territory (NT) in 2001 (Pavelic *et al.*, 2002b). An infiltration gallery for stormwater recharge was established at Kensington New South Wales in 2007. Soil aquifer treatment of reclaimed water began in Alice Springs NT in 2008. Investigations for stormwater ASR are underway at several sites in Melbourne, Victoria in 2008.

Stormwater disposal wells in Mt Gambier, SA, that have operated since the 1880s, were proven in the 1990s to contribute to the city's water supply drawn from Blue Lake. Subsequently Wolf *et al.* (2006) have established risk management plans that, on being adopted, turned unmanaged recharge to MAR. In Perth, recharge of roof runoff and stormwater into pits and basins makes a substantial contribution to the water recovered from wells for household irrigation and also to a few mains water supply wells. Blanket risk assessments are proposed to convert this formerly unmanaged recharge to managed aquifer recharge.

1.5 Purposes of MAR

In rural areas, managed aquifer recharge has been used primarily for increasing the security of groundwater irrigation supplies and improving the quality of irrigation water. In urban areas there can be many other drivers for managed aquifer recharge (see Figure 1.6). For example, in Salisbury, ASR was really only considered after wetlands were established for flood mitigation, urban amenity and coastal water quality improvement. The additional costs of injecting and recovering the detained water provided a water supply competitive with mains water prices.

For new projects, there may be multiple reasons for introducing MAR at any one site. The combination of benefits as stated below, not just water supply, may determine whether a MAR project proceeds.

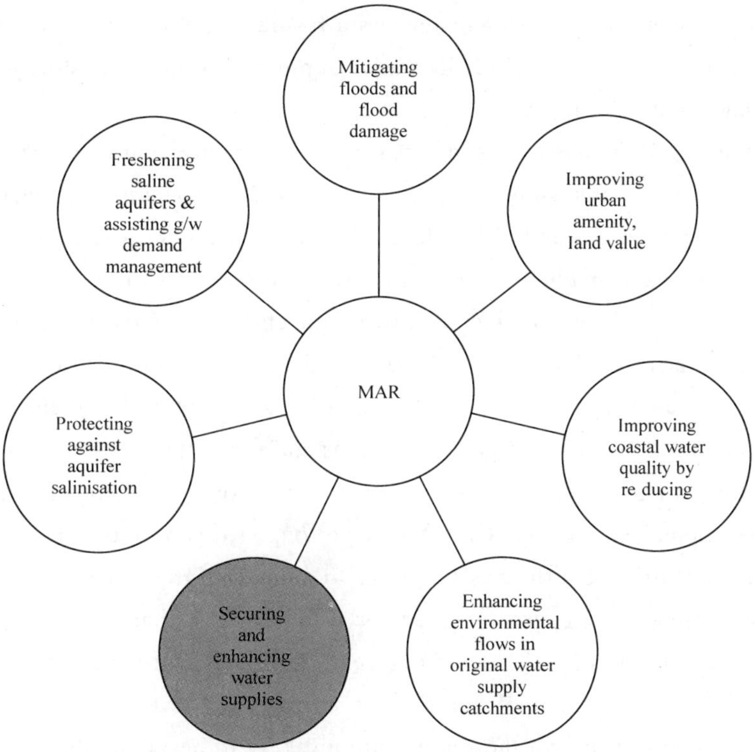

Figure 1.6 Managed aquifer recharge projects, particularly in urban areas, have many objectives in addition to water supply, and these vary from site to site. All of the economic, social and environmental benefits and costs of projects need to be taken into account in project selection.

In the specific case of securing and enhancing water supplies, pioneering Australian MAR projects were established to satisfy immediate non-potable needs on a commercial basis for local government, sporting clubs and industry.

Application of MAR for establishing high-valued drought and emergency supplies could increase as governments and water utilities become more confident of MAR costs and reliability in relation to alternatives and when multi-purpose benefits and costs are taken into account. Emergency supplies have highest social value but unless procedures are established to support this investment and to protect strategic reserves, aquifer storage capacity in urban areas will be consumed for commercial purposes. A hierarchical approach is needed to allocate aquifer storage space and harvestable stormwater, first to strategic reserves and substitutional supplies. Residual capacity, if any, could be allocated to new supplies. This would help optimise the value of these resources to city water security, especially in light of climate change and population growth.

1.6 Climate variability as a driver for MAR

In the future, much of the continent will face decreasing mean annual rainfall, shifts in seasonal patterns, more frequent high intensity rainfall, higher temperatures and higher evaporation rates because of climate change. As a result, to retain the current level of urban water security will require more stable alternative supplies, larger water storages, and a range of demand management measures.

Climate change will also affect land use, soil cover, erosion and fire frequency, so water quality from traditional water supply catchments is likely to become increasingly variable. Water treatment may be needed. Therefore, the water quality advantages of traditional catchments over urban catchments are likely to partially erode over time.

Bank filtration could be adopted by towns currently using river water directly, as a measure to adapt to climate change. This MAR practice is common in Europe where drinking water supplies are drawn from wells in alluvium next to streams rather than the stream itself as a means of smoothing out water quality variations and pre-filtering water.

The need for storage depends to a large extent on seasonal variations in water sources in addition to inter-annual variability. Cities with prolonged dry periods (in Figure 1.7, cities with low fractions of driest continuous 6 months to annual rainfall, or with low mean annual rainfall to evaporation) have a greater need for water storage than wetter cities with more uniform rainfall (upper right hand side of Figure 1.7). However, for very dry cities (extreme left hand side of Figure 1.7) the opportunities for stormwater harvesting are infrequent and so MAR projects with stormwater in these areas are unlikely to be viable due to the low rate of utilization of the invested capital. In these locations reclaimed water, which is a very stable flow, is likely to be the preferred source of water for MAR.

In wetter cities there is more stormwater available, and therefore the unit costs of harvesting are reduced. So from a climate perspective the Australian cities of Darwin, Brisbane, Sydney and Hobart, which look more promising for stormwater harvesting, surprisingly have been less active than Adelaide, Perth and Melbourne in advancing MAR. Reasons for this in part may relate to the presumed aquifer properties beneath these cities, a topic which will be discussed later.

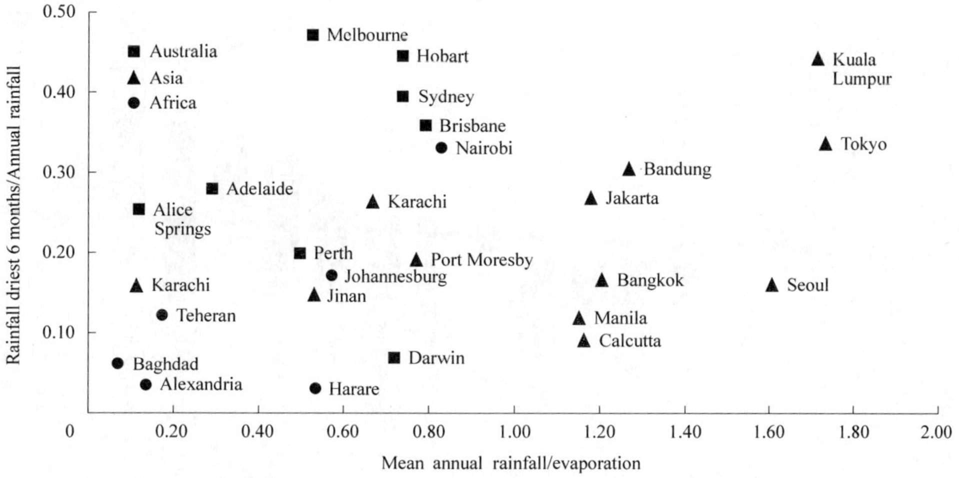

Figure 1.7 Seasonality of rainfall, shown as the ratio of driest continuous 6 months to annual rainfall, and ratio of rainfall to evaporation, are indicators of the need for inter-season water storage and the likely availability of stormwater to meet water demand.

1.7 Costs of MAR with respect to alternative water supplies

Managed Aquifer Recharge is commonly applied as part of water recycling for indirect potable or non-potable use of stormwater or reclaimed water to substitute for all or some uses of existing mains water supplies. In any city, town or rural area, the costs and benefits of MAR can be compared with a range of options including improved water conservation, tapping new surface water supplies or aquifers, rainwater tanks, and groundwater or sea water desalination. The local situation dictates the costs of each option and large variations occur between localities for costs of any option and therefore the relative cost effectiveness of different options.

Figure 1.8 demonstrates the range of unit costs evaluated on a consistent basis across four cities for a range of investment options. MAR may play a role in storage, additional passive treatment and distribution for 3 options shown(stormwater reuse, indirect potable reuse and non-potable recycling).

Most cities will require diversified sources of water to enhance security of supply, taking account of the size, quality, reliability, proximity and cost of each new resource or water conservation measure. Before tapping new sources, investment is needed to assess the attributes of each resource and develop plans for an integrated portfolio of measures to secure future water supplies that account for climate change, population growth, changing demographics, existing infrastructure and its longevity, as well as environmental and social measures.

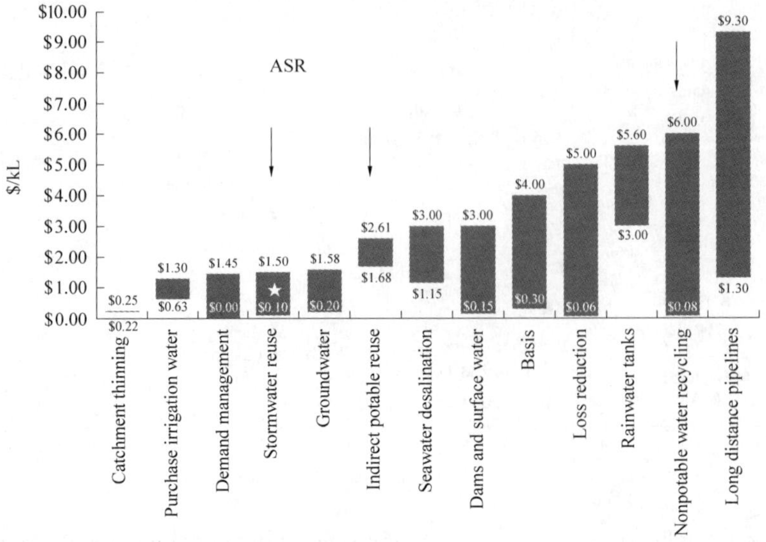

(Source: Marsden Jacob Analysis based on water supply plans for Sydney, Adelaide, Perth, Newcastle. Lower bound of indirect potable reuse estimate based on Toowoomba.)

Figure 1.8 Comparison of direct costs of water supply enhancement and demand reduction options for four Australian cities. (Source: PMSEIC, 2007, Water for Our Cities) Blue bars represent the range in costs for each option.

Figure 1.9 shows an example of the range of options available to a city taking account of potentially reduced security of supplies from established resources and demand growth. Wedges represent options that could potentially be developed to contribute to the portfolio of water supply capabilities. The extent to which each option will ultimately contribute depends on the estimated sizes of each alternative resource and the financial costs of developing it. For example, seawater desalination is not capacity constrained nor rainfall dependent and can be built relatively quickly, but it is the most expensive option for supply development. Hence

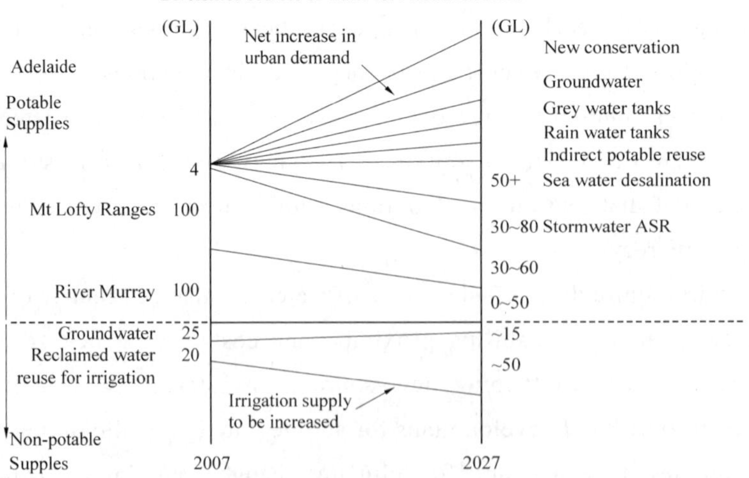

Figure 1.9 Diversified sources of water may include MAR as one of the elements of future city supplies that may complement other options such as desalination, and expand storage and robustness of supplies.

expansion beyond immediate needs would depend on any remaining gap between cheaper sources of supply and demand.

Stormwater ASR/ASTR is a low cost option for potable supplies as well as irrigation and industrial supplies to replace the use of potable water for non-potable uses. Reclaimed water ASR will be needed to increase the amount of reclaimed water available for horticulture, viticulture and non-potable urban uses.

1.8 Situations where MAR is not viable

Managed aquifer recharge is only feasible if there is a suitable aquifer; that is, one that can accept a sufficient volume of water at a sufficient recharge rate for the benefits to justify the costs of establishing the project. If no suitable aquifer is detected within the affordable drilling depth, and local hydrogeological studies do not reveal a suitable aquifer near the source or sufficient demand for water to allow its economical transfer, then managed aquifer recharge is not feasible.

If well yields are low, viability can be determined by calculating a minimum economic rate of recharge or recovery. The number of injection or recovery wells, or the area of infiltration galleries or ponds required to recharge and recover water at the required rate can then be compared to the project's budget. Generally, if the yield of wells in an aquifer is marginal, then treatment for recharge water to avert clogging will generally have a high cost; alternatively, maintenance costs will be even higher per unit volume of water recovered.

Managed aquifer recharge is not recommended if environmental risk cannot be reduced to an acceptably low level by economically viable preventive measures (taking all costs and benefits of the project into account). Marginally feasible projects — those with only small net benefits to the proponent—are not encouraged if incentives and capacity for effective management are low.

An unconfined aquifer with a shallow watertable also presents problems. There is nowhere to store additional recharge without causing waterlogging, salinisation, or geotechnical problems for buildings and other infrastructure. Unless the managed aquifer recharge project is an integral part of a proposal for increased extraction of water, such sites should be avoided. However, management of these situations is highly constrained and watertable fluctuations are likely to significantly exceed those that occur naturally.

If the available storage capacity of an aquifer is already fully committed to other managed aquifer recharge operations, then additional managed aquifer recharge is impractical and would have a negative effect on the performance of the existing operations.

Managed aquifer recharge also requires a source of recharge water for sufficient time to ensure that the recoverable volume warrants the project's establishment costs. Urban stormwater requires land for detention storage. In new subdivisions, the increase in land value due to water views exceeds the costs of land used for detention storage. In built-up areas, finding land for detention storage is much more difficult, unless managed aquifer recharge is a coherent component of an urban renewal project.

2 How to Establish a MAR Project Using a Risk Assessment Framework

Potential proponents need to know first whether they have the five essential elements (Stage 1 assessment) of a MAR project outlined below before proceeding further. If the project looks potentially viable at this first stage, the Draft MAR Guidelines (EPHC-NHMRC-NRMMC, 2008b) lead proponents through the investigations (Stage 2) and commissioning trials (Stage 3) to an operational project (Stage 4).

2.1 Stage 1 — Five essential ingredients

The five critical elements for a successful MAR project (see Figure 2.1) are:

- a sufficient demand for recovered water;
- an adequate source of water for recharge;
- a suitable aquifer in which to store and recover the water;
- sufficient land to harvest and treat water;
- capability to effectively manage a project.

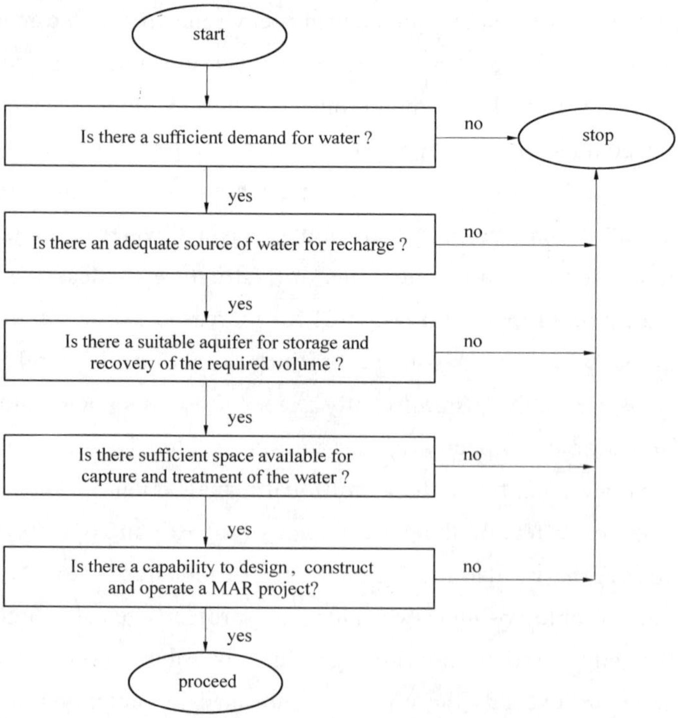

Figure 2.1 A checklist for considering whether to undertake a managed aquifer recharge project

(1)Demand: The volumetric demand for recovered water (within an economic scale) or a clearly defined environmental benefit of recharge is essential for MAR. The purposes for which water will be recovered also need to be defined (see Figure 1.4). Generally this will provide the revenue stream to pay for the water supply cost elements of the project. In urban areas demand for stormwater detention for mitigating floods, improving coastal or receiving water quality and enhancing urban amenity and land value may also contribute revenue streams for MAR projects. For reclaimed water projects the decline in discharge of treated effluent to sea may provide a motivation for investment in MAR.

(2)Source: Entitlement to water to be used for recharge needs to be secured. Mean annual volume of recharge should exceed mean annual demand with sufficient excesses to build up a buffer storage to meet reliability and quality requirements. In an over-allocated catchment it is unlikely that an entitlement to surface water would be available.

Stormwater and reclaimed water are usually abundant resources in urban areas but require treatment and storage before reuse. Mean annual stormwater discharge is between 85% and 145% of mains water use and sewage effluent discharge is between 50% and 80% of mains water use across Australian cities.

The availability of stormwater or reclaimed water to make useful contributions to city water supplies is not a constraint. The primary limitation to stormwater harvesting and use is the ability to store the water from runoff events for subsequent use as drinking water supplies or as irrigation, industrial supplies or other non-drinking uses. MAR can provide an economic means of storing water in urban areas.

Sewage effluent requires extensive treatment before placement in either dams or aquifers prior to reuse. Aquifers have advantages with respect to ongoing passive treatment of the water and allowing longer assured residence times before recovery for drinking.

(3)Aquifer: A suitable aquifer is critical for MAR. It needs to have an adequate rate of recharge, sufficient storage capacity and be capable of retaining the water where it can be recovered. Low salinity and marginally brackish aquifers are preferred so that mixing with fresh recharge water should still allow recovered water to be fit for use.

Local hydrogeological knowledge is needed to identify the presence of aquifers and their suitability for MAR. Basic stratigraphic and hydrogeological information existing wells can serve as valuable background information before drilling new wells and can be used to determine potential sites for MAR (see Figure 2.2). Hydrogeological reports generally provide some indication of the level of knowledge of the local aquifers and their degree of uniformity. Aquifer properties vary spatially so it is not generally reliable to extrapolate from one site to predict viability or performance at a nearby site.

At any given location there may be several aquifers stacked on top of each other interleaved with low permeability layers. This allows choice of one or more with the most favourable characteristics for water storage. Depending on their degree of inter-connection, it may be possible to store water of different qualities in different aquifers at the same location.

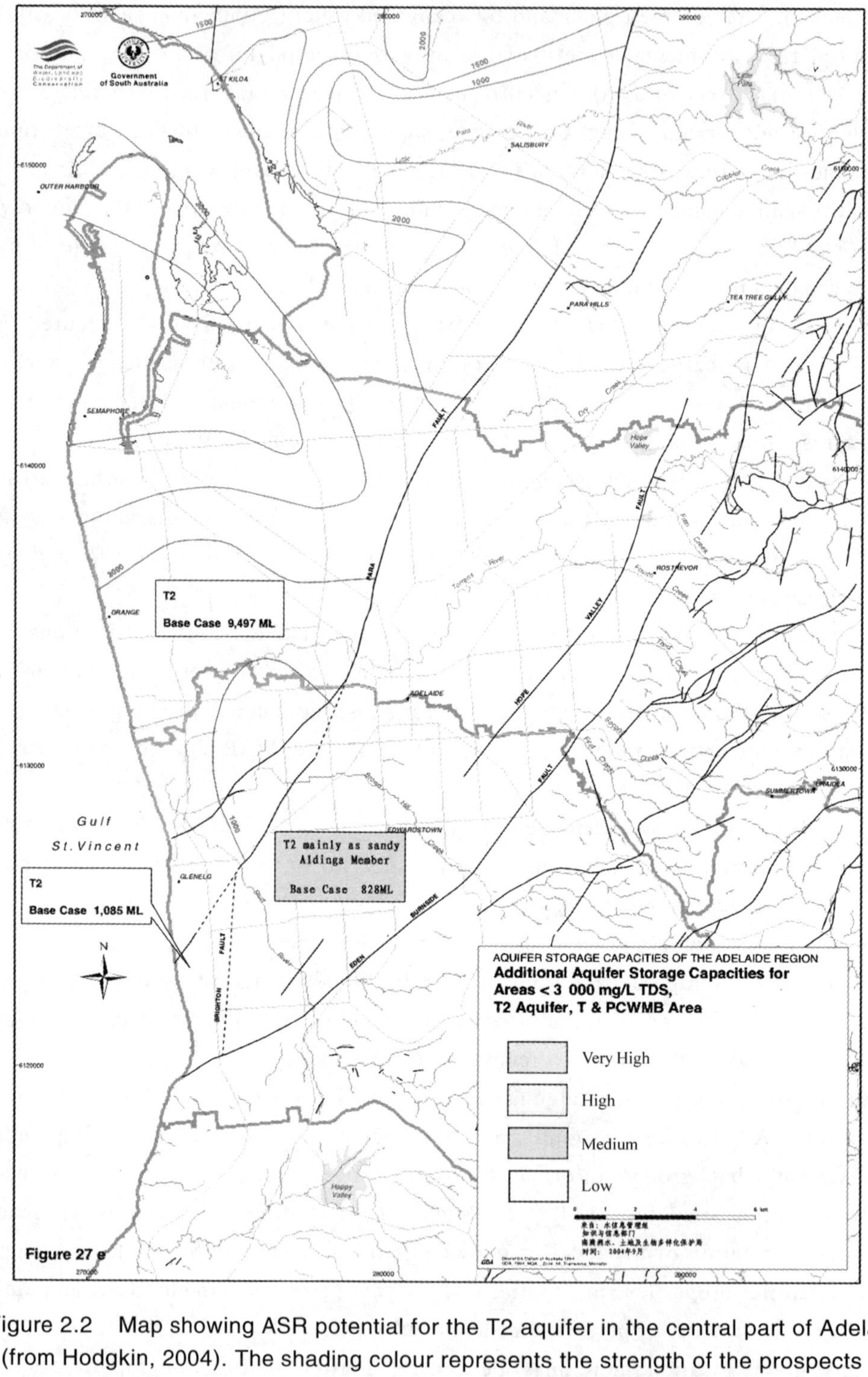

Figure 2.2　Map showing ASR potential for the T2 aquifer in the central part of Adelaide (from Hodgkin, 2004). The shading colour represents the strength of the prospects for ASR projects, based exclusively on the hydrogeology of the areas. The dominant factor is the potential rate of injection considered to be achievable by Interpolation from well records and historical hydrogeological reports

(4)Detention Storage: There should be open space, or dams, wetlands, ponds or basins to detain sufficient water without causing flood damage to enable the target volume of recharge to be achieved. Similarly there needs to be space available for whatever treatment process, if any, is subsequently determined to be required. In established urban areas space for capture can be a major impediment to stormwater water harvesting and ASR wells are commonly used to avoid land requirements of infiltration systems. For recycled water from a sewage treatment plant generally no additional detention storage will be required at the recharge facility.

(5)Management capability: Hydrogeological and geotechnical knowledge, as well as knowledge on water storage and treatment design, water quality management, water sensitive urban design, hydrology and modelling, monitoring and reporting are all required to meet governance requirements. Such expertise will be required from Stage 2 and a growing number of consultants are experienced in investigations and design of MAR projects.

2.1.1 Identify the degree of difficulty

Table 2.1 gives an example of a checklist to understand the degree of difficulty associated with a conceived project. This serves as a guide to the amount of effort required in project investigations and commissioning trials in order to manage human health and environmental risks in accordance with the relevant water quality management guideline (for example, National Water Quality Management Strategy). This consists of 13 questions with commentary on information required to answer and the consequences of the answer on the need for further information during the investigation stage (Stage 2).

Table 2.1 Example of assessment to determine the degree of difficulty of the conceived MAR project

Information required for assessment		Questions and indicators of degree of difficulty
1.Source water quality with respect to groundwater environmental values	Where multiple samples are available, the highest concentration of each analyte should be used in the evaluation; unless there is justification that events resulting in those values will be prevented when the MAR project is established Alternatively, in the absence of water quality data from actual source water, data may be used from existing similar MAR projects using the same type of source water and recharging the same aquifer In the absence of either data source above, generic data may be used Assessment of quality variability and factors affecting quality are deferred to the maximal risk assessment	Q1. Does source water meet the water quality requirements for the environmental value of ambient groundwater? If Yes—low risk of pollution is expected. This is a necessary condition, but not a sufficient condition for low risk If No—high maximal risk is likely. Expect Stage 2 investigations to assess preventive measures to reduce risk of groundwater contamination beyond attenuation zone (and size of attenuation zone)

Table 2.1 Continuous

Information required for assessment		Questions and indicators of degree of difficulty
2. Source water quality with respect to recovered water end use environmental values	If the source water does not meet the water quality requirements for the environmental values of intended end uses of recovered water, then there is a reliance on attenuation of hazards within the subsurface	Q2. Does source water meet the water quality requirements for the environmental values of intended end uses of water on recovery? If Yes—low risk of pollution of recovered water is expected. However, this is not a sufficient condition for low risk due to aquifer reactions If No—high maximal risk is likely. Expect Stage 2 investigations to assess this risk
3. Source water quality with respect to clogging	Where source water quality is poor and soil or aquifer are fine-grained, clogging of the infiltration basin and gallery or recharge well is likely to occur, unless the water is pretreated before recharge Clogging is most prevalent when water contains moderate or high levels of suspended solids or nutrients, such as nitrogen or labile organic carbon Clogging can also occur when oxygenated water is introduced into an aquifer that contains iron. If the soil or aquifer are coarse grained or contain macropores, clogging with such waters is less likely, but the risk of pollution of groundwater is high (as covered in Q1 and Q2) Lack of evidence of clogging is insufficient to indicate that risk of pollution is low, even in fine-grained media	Q3. Does source water have low quality, for example: total suspended solids >10 mg/L total organic carbon >10 mg/L total nitrogen >10 mg/L? Is soil or aquifer free of macropores? If Yes—high risk of clogging of infiltration facilities or recharge wells. Pretreatment will need consideration regardless of answers to Q1 and Q2 If No—lower risk of clogging is expected. However, this is not a sufficient condition for low risk, due to dependence of clogging on aquifer characteristics that would be revealed by stage 2 investigations
4. Groundwater quality with respect to recovered water end use environmental values	Where samples are available, the highest parameters detected in each sample should be used in the analysis; unless there is justification that events resulting in those values will be prevented when the MAR project is established Alternatively, in the absence of data on groundwater quality from the proposed site, data from nearby wells in the same aquifer may be used	Q4. Does ambient groundwater meet the water quality requirements for the environmental values of intended end uses of water on recovery? If Yes—low risk of inadequate recovery efficiency is expected If No—some risk of inadequate recovery efficiency is expected

Table 2.1 Continuous

Information required for assessment		Questions and indicators of degree of difficulty
5. Groundwater and drinking water quality	The environmental values of the aquifer need to be defined by the relevant authority. These will depend on the ambient groundwater quality and any groundwater affected ecosystems Setting these values involves a stakeholder consultation process, and in practice will possibly be related to groundwater allocation planning processes	Q5. Is either drinking water supply, or protection of aquatic ecosystems with high conservation or ecological values, an environmental value of the target aquifer? If Yes—high risk of groundwater pollution if recharged by water if answer to Q2 is No If No—low risk of groundwater pollution is expected. However, this is not a sufficient condition for low risk
6. Groundwater salinity and recovery efficiency	If native groundwater has high salinity, its proportion that can be present as a mixture with source water in recovered water is limited At such sites, density affected flow may also occur. Fresh recharge water can form a lens above the native saline groundwater, making recovery difficult and reducing recovery efficiency (ie. the volume of recovered water meeting the environmental values for its intended uses as a proportion of the volume of recharged water)	Q6. Does the salinity of native groundwater exceed (a) 10 000 mg/L or (b) the salinity criterion for uses of recovered water? If Yes to both—high risk of achieving only low recovery efficiency. Aquifer hydraulic characteristics, especially layering within the aquifer will need careful examination in Stage 2 If Yes to only (b)—moderate risk of low recovery efficiency is expected. However, this is not a sufficient condition for low risk (eg in brackish aquifers with high rates of ambient lateral flow) If No to both—low risk of low recovery efficiency
7. Reactions between source water and aquifer	Reactions between source water and aquifer minerals may result in deterioration of water quality for recovered water, and possibly for water in the aquifer beyond the attenuation zone; or cause excessive clogging or dissolution of the aquifer A full evaluation may be undertaken in Stage 2, but a simple indicator of the likelihood of potential problems at entry-level stage is to note the extent of contrasts between quality of source water and native groundwater	Q7. Is redox status, pH, temperature, nutrient status and ionic strength of groundwater similar to that of source water? If Yes—low risk of adverse reactions between source water and aquifer is expected. However, this is not a sufficient condition for low risk If No—high risk of adverse reactions between source water and the aquifer is possible, and will warrant geochemical modelling in Stage 2

Table 2.1　Continuous

Information required for assessment		Questions and indicators of degree of difficulty
8. Proximity of nearest existing groundwater users, connected ecosystems and property boundaries	Proximity of nearest existing groundwater users and groundwater-connected ecosystems is likely to influence the extent of investigations required in Stage 2 Typically, attenuation zones will have aquifer residence times of up to a year If property boundaries are close to the MAR site, then the attenuation zone may extend beneath a neighbouring property Groundwater pressure effects in confined aquifers due to MAR may propagate over considerably longer distances than water quality effects	Q8. Are there other groundwater users, groundwater–connected ecosytems or a property boundary near (within 100–1000 m) the MAR site? If Yes—high risk of impacts on users or ecosystems is possible, and this will warrant attention in Stage 2 If No—low risk of impacts on users or ecosystems is likely. However, this is not a sufficient condition for low risk
9. Aquifer capacity and groundwater levels	Groundwater mound height induced by MAR depends on aquifer hydraulic properties, size of recharge area and recharge rate Mounding is normally calculated in Stage 2 when aquifer properties are measured. However, excessive mounding can cause: Waterlogging; soil heave; flooding of below-ground infrastructure; salt damp; soil salinisation Hence, unconfined aquifers with shallow watertable sites are generally unsuitable as storage targets for large-scale recharge projects For confined artesian aquifers, care needs to be taken against overpressurisation, and to seal existing wells that might otherwise start to flow	Q9. Is the aquifer confined and not artesian, or is it unconfined, with a watertable deeper than 4 m in rural areas or 8 m in urban areas? If Yes—low risk of water logging or excessive groundwater mound height is expected. However, this is neither a necessary nor a sufficient condition for low risk If No—high risk of water logging or excessive groundwater mound height is expected. However, Stage 2 investigations may reveal that risk is acceptable
10. Protection of water quality in unconfined aquifers	If the aquifer is unconfined and the intended recovery is for drinking water supplies, then overlying land and waste disposal (including intensive horticulture and septic tanks) should be managed carefully or precluded from the groundwater capture zone	Q10. Is the aquifer unconfined, with an intended use of recovered water that includes drinking water supplies? If Yes—high risk of groundwater contamination from land and waste management If No—lower risk of groundwater contamination from land and waste management

Table 2.1 Continuous

Information required for assessment		Questions and indicators of degree of difficulty
11. Fractured rock, karstic or reactive aquifers	If the aquifer is fractured rock or karstic, the ability to recover stored water will require evaluation, especially if the ambient groundwater is saline, or hydraulic gradient is steep Provision will also need to be made for a larger attenuation zone, due to more rapid migration of recharge water from the recharge area	Q11. Is the aquifer composed of fractured rock or karstic media, or known to contain reactive minerals? If Yes—high risk of migration of recharge water is expected. There is a need for an enlarged attenuation zone, beyond which preexisting environmental values of the aquifer are to be met. Dissolution of aquifer matrix and potential for mobilisation of metals warrant investigation in Stage 2 If No—low risk of the above is expected. However, this is not a sufficient condition for low risk
12. Similarity to successful projects	A founding principle of MAR is that all validation and verification monitoring data should be in the public domain, and include sufficient operational data to enable accurate interpretation This information is of value for future MAR projects, for improving design and operation and reducing costs and further refinement of these guidelines A national or state repository for these data should be accessible for proponents	Q12. Has another project in the same aquifer with similar source water been operating successfully for at least 12 months? If Yes—take validation and verification data from the existing project(s) into account when designing the current project and the Stage 2 investigations and subsequent risk assessments If No—expect that all uncertainties will need to be addressed in the Stage 2 investigations
13. Management capability	A proponent new to MAR operation needs to gain appropriate expertise in parallel with Stage 2 investigations, to demonstrate a low level of residual risk for the precommissioning risk assessment.	Q13. Does the proponent have experience with operating MAR sites with the same or higher degree of difficulty, or with water treatment or water supply operations involving a structured approach to water quality risk management? If Yes—low risk of water quality failure due to operator experience If No—high risk of water quality failure due to operator inexperience. The proponent is recommended to gain instruction in operating such systems (eg a MAR operator's course or aquifer storage and recovery course) or engage a suitable manager committed to effective risk management in parallel with Stage 2, to reduce precommissioning residual risks to low

The questions address the water quality of the source water in relation to environmental values of the aquifer, of intended uses of recovered water, potential for clogging and potential for mineral reactions. They ask about groundwater quality in relation to recovered water uses, and whether groundwater needs to be protected for drinking supplies or high conservation ecological values, or is highly saline. They also ask whether there are nearby groundwater users or ecosystems, is the aquifer confined or artesian, fractured or cavernous, are if there similar projects with similar source water in the same aquifer and whether the proponent or his consultant or intended operator has experience with MAR or water quality management.

Costs of MAR investigations and trials are not trivial and having completed this checklist the proponent should know whether their proposed project has a low or high degree of difficulty and the types of information which will be of most value in the investigation stage. Because of the costs of these investigations it is normal to first seek assurance that at least the core approvals for MAR (see 2.1.2 section) are likely to be obtained, before investing in such investigations.

2.1.2 Approvals required

Currently there is no unified process for MAR project approvals in any Australian state or territory. Approvals that may be needed for a MAR project to proceed include:

- an entitlement to a share of the source water, such as stormwater, reclaimed water or other source, taking account of environmental flows and other users of the source water;
- a permit to construct wells for investigations, ASR or recovery;
- planning approval for a water impoundment, covering geotechnical safety, amenity, insect and pest nuisance and danger of drowning;
- a declaration of environmental values of an aquifer, accounting for ambient groundwater quality and current uses;
- approval to recharge water to an aquifer, to protect an aquifer's environmental values, prevent excessive changes in the hydraulic head, and to protect human health and the environment as a result of the recovery of stored water for intended uses;
- an entitlement to a share of the aquifer storage space, recognising that this is finite, and may be smaller than the harvestable volume of source water;
- an entitlement to recover water from an aquifer, possibly as a proportion of the cumulative recharge that may depend on the degree to which the aquifer is over-exploited and will take account of other groundwater users and water bankers so as not to cause them adverse impacts;
- transfer of water entitlement, endowing the recharger with an ability to transfer their entitlement to a third party, subject to hydrogeological constraints and not into locations where piezometric heads are already depressed;
- a permit for the use of recovered water, to ensure that usage conforms with catchment management plans and that the water quality is fit for intended uses.

Water resources management authorities at state and/or regional level are normally responsible for water quantity entitlements; environment protection agencies are normally responsible for water quality and approvals to recharge; local government is responsible for planning and development approvals; health commissions are responsible for permits for use; and water utilities will also be involved if reclaimed water is a source or their drinking water mains or non-potable supply mains are used to distribute recovered water.

2.2 Next steps

Assuming that the entry level assessment (Stage 1) indicates that the project is apparently viable, the degree of difficulty does not deter the proponent, and the regulator has not identified other impediments, the next stage is to undertake investigations on source water, pretreatment methods and the aquifer to determine if the project will demonstrably protect human health and the environment, notably the aquifer.

Stage 1 was a rapid qualitative assessment but Stage 2 is quantitative, using existing information supplemented by site-specific investigations that were foreshadowed in Stage 1. Information to confirm that the project is operating as intended will not be available until commissioning of a pilot project, or the full-scale project after it is constructed. A staged approach (see Figure 2.3) to project development helps avoid wasting time and money, and can improve the design of the project by tailoring it to the aquifer. It also allows investment appropriate to the MAR project in relation to alternative or complementary projects.

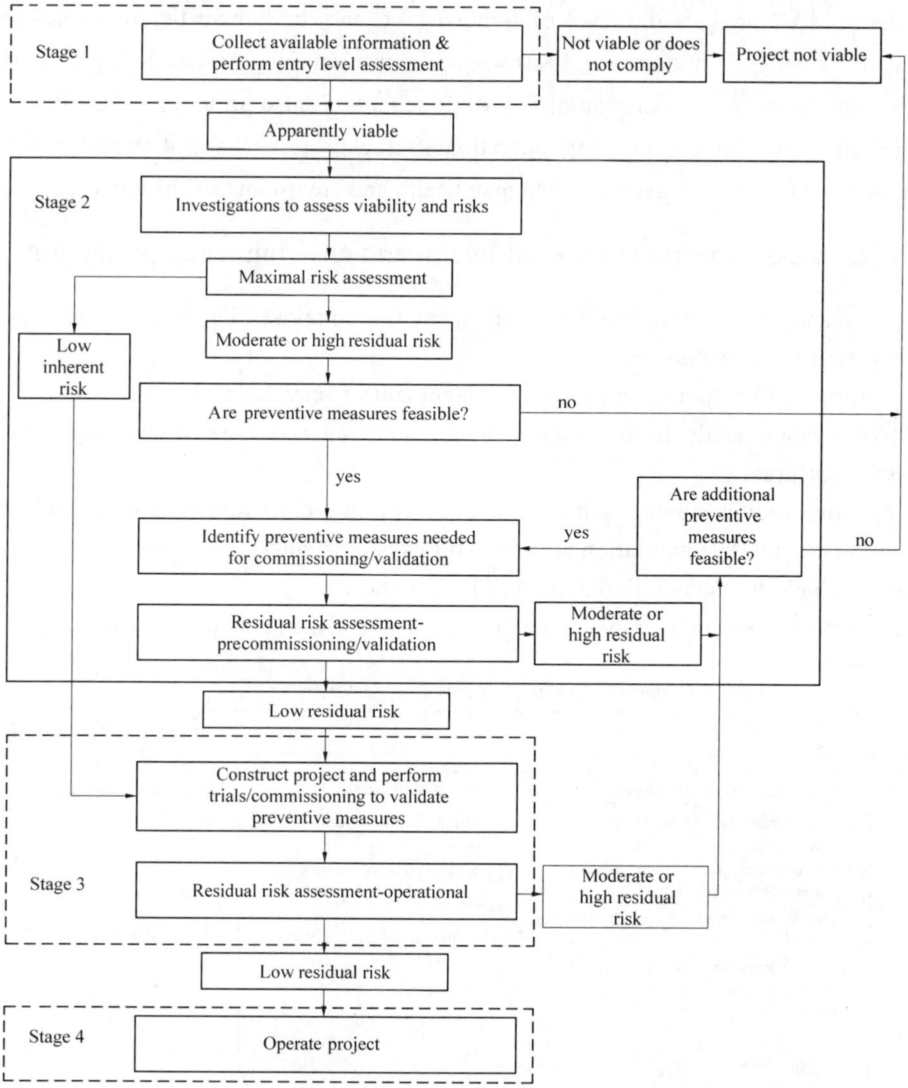

Figure 2.3 Stages in establishing a MAR project to meet human health and environmental needs (figure from Peter Newland, SA EPA)

Stage 2 investigations enable risks to be assessed and the preventative measures by which they can be managed to be indentified. This requires information describing the source water quality, infrastructure and proposed operations of the project, and characterisation of the hydrogeology to demonstrate that all hazards have been addressed with sufficient supporting information for a management plan. Where further information is needed, a pilot project might be required.

At this point the risks of success will be better defined and a decision would be made on whether to invest in constructing the MAR project or in alternative water supplies. Such a decision would account for the full range of costs and benefits of all projects.

For many MAR projects the level of some risks cannot be known before implementation, accompanied by suitable monitoring. Known as commissioning trials, Stage 3 monitoring provides a basis for ongoing operation of the MAR scheme. It also provides for the development of management plans for the ongoing operation of the project, Stage 4, which will require ongoing monitoring to ensure that risks to human health and environmental health are controlled.

2.3　A risk assessment framework for health and environmental protection

The risk management framework for MAR operations comprises 12 elements (see Figure 2.4), that fall into four main categories:

- commitment to responsible use and management of recycled water;
- MAR system analysis and management, such as risk assessment and a series of preventative measures;
- supporting requirements, such as employee training, community involvement, research and development, and documentation and reporting systems; and
- review, including evaluation and audit processes.

All 12 elements need to be implemented for the risk management approach to be successful.

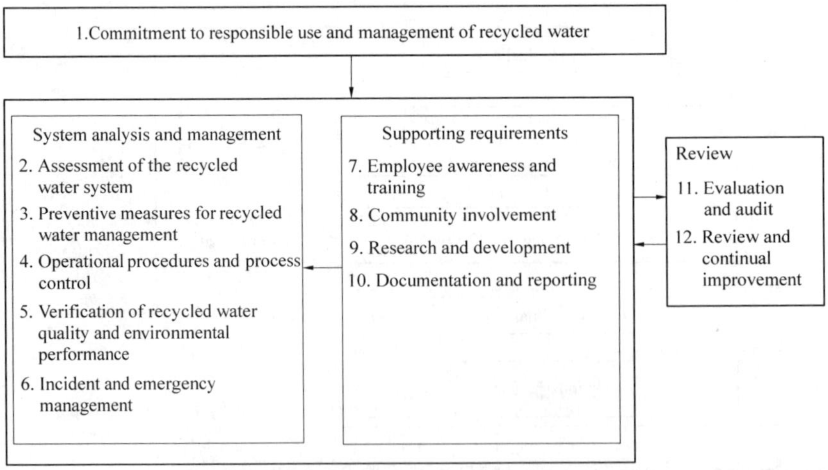

Figure 2.4　Elements of the framework for managing water quality and use.
(EPHC-NHMRC-NRMMC, 2008b)

This approach aims to protect the environmental values of all intended uses of recovered water and of the aquifer beyond a transient attenuation zone, and to prevent adverse impacts. This is done by assessing potential hazards and the risks associated with each, and identifying preventive measures to manage the risks.

A simplistic view that treating water to near drinking standards before recharge will protect the aquifer and recovered water is incorrect. For example chlorination, to remove pathogens that would sustainably be removed in the aquifer, can result in water recovered from some aquifers containing excessive chloroform. In some locations, drinking water injected into potable aquifers has resulted in excessive arsenic concentrations on recovery due to reactions with pyrite containing arsenic. Source water that has been desalinated to a high purity dissolves more minerals within the aquifer than water that has been less treated.

Hence a scientific approach is recommended that takes account of three ways that aquifers interact with recharged water:

(1)Sustainable hazard removal—allowing for attenuation during passage through soil and aquifer within an attenuation zone (see Figure 2.5), (eg pathogen inactivation, biodegradation of some trace organics, a limited amount of nutrient assimilation).

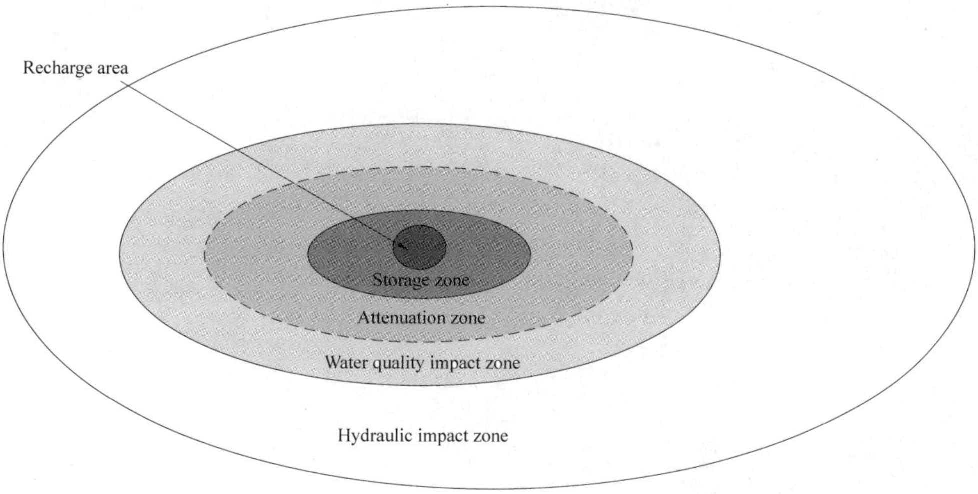

Figure 2.5 Schematic showing zones of influence of a MAR operation

(2)Ineffective hazard removal—these hazards need to be removed prior to recharge because they are either unremoved (eg salinity) or removal is unsustainable (eg adsorption of any metals and organics that are not subsequently biodegraded, or excessive nutrients or suspended solids).

(3)New hazards introduced by aquifer interaction (eg metal mobilization, hydrogen sulphide, salinity, sodicity, hardness, or radionuclides)—there is a need to change the quality of recharge water to avoid these (eg change acidity—alkalinity, reduction-oxidation status or reduce nutrients).

The response of an aquifer to any water quality hazard depends on specific conditions within the aquifer including temperature, presence of oxygen, nitrate, organic carbon and other nutrients and minerals, and prior exposure to the hazard.

The zone of aquifer in which water quality may be measurably affected by MAR may be larger than the attenuation zone, but in this outer domain the water quality should continuously satisfy the initial environmental values of the aquifer The effects of MAR operations on hydraulic heads (pressures) may be measurable over a much larger area, especially in confined aquifers, and may extend several kilometres. If the aquifer is originally too saline for the uses of recovered water, a storage zone can be identified that contains water which when recovered is fit for its intended use (see Figure 2.5).

The dotted line in Figure 2.5 marks the outer boundary of the attenuation zone. This represents the maximum separation distance between the MAR recharge structure and well(s) for verification monitoring to ensure that the ambient groundwater quality is protected. As the attenuation zone defines sustainable attenuation only, on cessation of the MAR operation this will shrink and disappear as ultimately the whole aquifer will meet all its initial environmental values.

3 Hydraulic Considerations

3.1 Clogging

Clogging is a reduction in the permeability of a porous medium. It is one of the most serious operational issues for MAR and affects most operations to some degree. Clogging is a more significant issue for well-injection systems, because the rate water enters the aquifer at is orders of magnitude higher than that for recharge basins. In turn, the rate in recharge basins is orders of magnitude higher than for natural recharge conditions.

Clogging develops during periods of recharge, as a result of interactions between the source water, its constituents and the porous media. It occurs mainly around the entry zone for recharge water, affecting the surface of infiltration systems or concentrating around the well-screen, gravel pack or adjacent aquifer in well-injection systems.

Clogging leads to a reduction in flow rates, thereby restricting the volume of water stored. This poses a constraint to the scale and economic viability of managed aquifer recharge operations, and in severe cases, can lead to the operational failure of managed aquifer recharge systems. Wells in inappropriately designed and operated managed aquifer recharge systems may clog within a matter of days or weeks; basins may clog within weeks or months.

Clogging occurs as a result of one or more physical, chemical or biological processes, such as:

- filtration of suspended solids;
- microbial growth;
- geochemical reactions;
- air entrapment and gaseous binding.

(1)Filtration of suspended solids. clogging of soil or aquifer pore spaces, by filtration of suspended solids present in the recharge water, results in the formation of a low-permeability clogging layer. Clogging is generally most severe where suspended solids concentrations are highest, but there are other determining factors. Significant factors are the flow rate, and hence the cumulative mass of sediment deposited within the porous media; and the hydrologic properties of the porous media that control the extent and penetration of deposition of particles within the system.

(2)Microbial growth. Microbial clogging of aquifers occurs through the multiplication of bacterial cells, and the production of extracellular polysaccharides (biofilms). Bacteria produce biofilms when nutrients (eg organic carbon, nitrogen, phosphorus) present in the recharge water stimulate microbial activity. Clogging due to growth of algae on basin floors can be an issue in surface infiltration systems. This may be compounded by precipitation of carbonate minerals, such as calcite, induced by photosynthetic consumption of carbon dioxide. Daily or seasonal variations in water temperature can affect the infiltration rate, through effects on algal growth and influence on the viscosity of the recharge water. All other factors being equal, infiltration and injection rates are typically higher in summer than winter.

(3)Geochemical reactions. Recharge of waters not in equilibrium with the groundwater or aquifer materials can cause chemical reactions. These lead to the production of insoluble precipitates that alter the permeability of the porous media. Such reactions may include:

- dissolution;
- precipitation;
- ion-exchange;
- ion-adsorption;
- oxidation-reduction.

If the aquifer contains iron-rich or manganese-rich minerals, then the presence of oxygen or nitrate in the recharge water can stimulate bacteria (eg Gallionella) to precipitate iron or manganese oxides and hydroxides that lead to clogging. Exchange of cations between the recharge water and the clays within the aquifer can lead to either swelling or dispersion. This tends to be most prevalent where reactive clays (eg montmorillonite) are present, and where there is a large decrease in the salinity of the recharge waters compared to the ambient groundwater. Other geochemical reactions, such as dissolution, may have an unclogging effect by increasing permeability (eg in places where calcite cement dissolves). In pronounced cases, this may lead to well instability. Precipitation of iron oxides and hydroxides and calcium carbonate are the predominant forms of geochemical clogging, but are not widely identified as the major clogging mechanism. This is because they coincide with other forms of clogging, or may take long periods of time to develop. Many geochemical reactions, especially redox reactions, are catalysed by bacteria; thus it is difficult to separate chemical and biological processes in many situations.

(4)Air entrapment and gaseous binding. Clogging by air entrainment can occur if water is allowed to cascade into the well, and bubbles are produced that block pore spaces and restrict flow. Dissolved gases may also be released from solution due to temperature changes, where cool source waters meet warm groundwater; or as a result of biogeochemical reactions, such as denitrification, which produces nitrogenous gases.

Further information on clogging in managed aquifer recharge systems can be found in Olsthoorn (1982), Pyne (1995), Pérez-Paricio and Carrera (1999), Bouwer (2002) and Pavelic *et al.* (2007).

3.1.1 Management of clogging

The generic measures of clogging are declining recharge rates and increased pressures. A summary of the methods used to diagnose and address the four main forms of clogging described above are presented in Table 3.1. This table indicates the means by which clogging may be minimised or prevented, which are:

Table 3.1 Clogging types, causes, diagnosis and methods for management

Clogging mechanism	Cause	Minimisation and prevention	Renovation technique	Measurements and techniques
Filtration of suspended solids	1. Particulates in source water or deposited in pipes 2.Treatment byproducts (eg alum) 3. Algal growth in basins	1. Improve pretreatment of source water, especially removal of suspended solids 2. Prevent light (algal growth) 3. Backwash wells or dry out basins more frequently	1. Pumping 2. Surging 3. Jetting 4. Basin drying or scraping	1. TSS, turbidity, MFI, particle size distribution of source water and soil/aquifer 2. Bypass filter testing 3. Changes in specific capacity (wells)
Microbial growth	Abundant substrate in source water (especially N, P and AOC)	1. Improve source water pretreatment, especially nutrient removal or chlorine dosing 2. Backwash more frequently 3. Adapt duration of wetting and drying cycles in basins	1. Pumping 2. Chlorine shock dosing 3. Acid treatment 4. Heat 5. Wire brushing 6.Multienzyme solutions 7.Basin drying or scraping	1. Water quality analyses 2. Photography (basins) and down-hole TV camera survey 3. Chemical and microbial analyses on soil and aquifer samples
Chemical precipitation, ion exchange	1. Redox reactions 2. Incompatibility between source water, groundwater and aquifer materials 3. Presence of reactive clays (swelling/dispersion)	1. Improve pretreatment 2. Adjust source water pH or cation concentration 3. Avoid or remove DO in source water 4. Backwash more frequently	1. Pumping 2. Wire brushing 3. Acid treatment 4. Basin scraping	1. Water quality analyses (especially redox conditions and mineral saturation indices) 2. SEM analysis of backwash water or soil and aquifer samples (scrapes of well face for deep systems) 3. Photography (basins) and down-hole TV camera survey
Air entrainment, gas binding	1. Free falling water 2. Presence of gas in source water 3. Biogeochemical reactions	1. Avoid cascading 2. Gas trap 3. Degas source water 4. Prevent reactions (eg N_2 production)	1. Pumping 2. Waiting for gas to redissolve 3. Improving well design	Analyse water temperature and specific gases

Note: AOC = assimilable organic carbon; DO = dissolved oxygen.

Source: adapted from Pérez-Paricio and Carrera 1999, Pyne 1995.

- adequate source water pretreatment;
- appropriate site selection;
- construction of well-designed installations;
- implementation of appropriate operational controls, renovation practices and monitoring.

A variety of measures can prevent or minimise clogging, where this offers a more effective strategy than managing a higher degree of clogging. For example, pretreatment of source water to remove suspended solids, organic matter and nutrients reduces clogging, and hence the degree of renovation required. Some pretreatment measures may be effective in dealing with clogging, but have less desirable side-effects. For example, disinfecting the source water will reduce microbial activity and minimise bioclogging, but can lead to the production of disinfection byproducts in the aquifer. Preventing air entrainment during well injection is relatively straightforward to achieve, by placing the intake pipe below the static water levels and fitting a flow control valve to ensure a positive pressure is maintained.

The aim of renovation is to return the performance of the injection well or recharge basin to its prior state, by restoring the hydraulic properties of the clogged zone (also known as unclogging). Unclogging is relatively straightforward for recharge basins, since it simply relies upon regular drying out and occasional scraping away of clogging agents.

In soil aquifer treatment systems, for example, periods of infiltration lasting from several days to several weeks are followed by periods where the basins are allowed to drain and dry out over similar time scales, although these may need to be seasonally adjusted.

A variety of mechanical methods have been used to redevelop injection wells. These include backwash pumping (typically at a rate higher than recharge), surging (variable speed pumping) or jetting with compressed air (also known as air lifting). Redevelopment is undertaken periodically, often daily, depending on how quickly recharge rates decline. Typically, only a few percent of the recharge volume is foregone.

Chemical methods used include the addition of oxidising agents (eg chlorine), flocculants (eg calcium chloride), or, in cases of severe clogging, acids. Renovation measures are effective as a routine operational method for unclogging wells and basins, demonstrated by the fact that some sites have operated continually for decades.

3.1.2 Tools for predicting clogging

Numerous techniques have been developed to predict clogging. These range in complexity from the measurement of simple water quality indices through to sophisticated process-based modelling. A summary of the main techniques is given below.

3.1.2.1 Measures of recharge water quality parameters

Measures of recharge water quality parameters that are useful indicators of the clogging potential of managed aquifer recharge systems include:

(1)physical clogging:turbidity,total suspended solids,membrane filtration index.

(2)microbial clogging:total organic carbon,dissolved organic carbon,assimilable organic carbon.

Application of a single parameter can be problematic when multiple forms of clogging occur at the same time. Characterisation of a prospective source water's or target aquifer's clogging potential should include assessment of the physical, chemical and microbiological composition of the source water. The assessment needs to take temporal variability into account. The physiochemical and mineralogical properties of the soil and aquifer and the quality of the ambient groundwater would also need to be defined.

3.1.2.2 Laboratory column studies

Laboratory column studies using soil and aquifer core materials and source waters can identify the extent of clogging and define source water pretreatment needs (eg Rinck-Pfeiffer *et al.*, 2000; Wood *et al.*, 2005). However, due to spatial variability in subsurface properties, these small-scale tests only offer a guide, and can not always be relied upon to accurately predict field-scale performance.

3.1.2.3 Graphical techniques

Simple graphical techniques to predict clogging have been developed. Hutchinson and Randall (1995) developed type curves for broad-based prediction of the physical clogging rate of injection wells, using aquifer transmissivity and MFI data. Taylor and Jaffé (1990) determined the relative reduction in permeability from biomass levels within the porous medium, using experimental data from a laboratory column study.

3.1.2.4 Numerical or mass balance modelling

Sophisticated numerical modelling codes that couple flow and transport to bio-geochemical reactions for site-specific prediction of porosity and permeability changes have been developed and applied (Pérez-Paricio and Carrera, 1998). However, these codes are highly complex, data intensive and only suitable for research purposes at present. Alternatively, simpler mass balance modelling can be performed, to gauge the likely impact of physical, chemical or microbial processes on permeability (Pavelic *et al.*, 2007).

3.1.3 Recharge water quality requirements for sustained MAR operations

Decades of operational experience have shown that clogging prevention is a better option than renovation, particularly for well injection systems. Whilst various renovation measures exist that can yield excellent results (see Table 3.1), ensuring that recharge water meets the appropriate water quality target through adequate pretreatment is a key factor in ensuring successful and sustainable long-term managed aquifer recharge operations. Although all forms of clogging can be avoided through pretreatment, in practice there is a trade-off between the costs associated with pretreatment, and the degree of clogging that is acceptable in terms of the type and frequency of renovation that would be required. Definitive guideline values on the quality of water needed for sustainable Australian managed aquifer recharge operations are a highly desirable target. As yet, this target is unachieved, and remains an area of active research.

Managed aquifer recharge practitioners in the Netherlands have determined, through trial and error, that the suitability of source water for injection into sandy aquifers is defined by a membrane filtration index of< 3~5 s/L^2 and assimilable organic carbon of <10 μg/L, to manage physical and biological clogging respectively (Olsthoorn, 1982; Hijnen and van der Kooij, 1992). Such single guideline values are inappropriate for Australia, which has a greater diversity of source water types and hydrogeological conditions than The Netherlands. Pérez-Paricio and Carrera (1999) developed guidelines on source water quality in relation to likely degree for clogging of injection wells and surface basins. These guidelines offer a useful starting point, but do not consider the soil or aquifer conditions.

Clogging rates at 14 aquifer storage and recovery sites were classified by quality of the source water and type of aquifer. The results show that both factors may affect clogging rates (see Table 3.2. For aquifer storage and recovery projects in alluvial aquifers, source waters with total suspended solids and total organic carbon of less than 10 mg/L have lower rates of clogging than more turbid or nutrient-rich waters. For poor quality water, limestone aquifers clogged slowest, followed by alluvial aquifers; fractured rock aquifers clogged fastest. These results are not surprising, because unclogging in limestone aquifers is facilitated by matrix dissolution, and alluvial (sand and gravel) aquifers offer more flow paths than fractured rock aquifers. To meet operational requirements, site-specific evaluation to assess clogging potential and identify water quality targets suited to the aquifer is needed at all sites (Pavelic *et al.*, 2007).

Table 3.2 Rates of clogging at 14 aquifer storage and recovery sites according to aquifer type and quality of source water

Aquifer type	Different source water quality of the rate of clogging		
	Low (TSS >10 mg/L; TOC >10 mg/L)	Moderate (TSS is 1 ~ 10 mg/L; TOC is 1 ~ 10 mg/L)	High (TSS <1 mg/L; TOC <1 mg/L)
Limestone	Low→moderate	Low→moderate	No data available
Sand and gravel	Moderate→severe	Low→moderate	Low→moderate
Fractured rock	Severe	No data available	No data available

Note: TSS = total suspended solids; TOC = total organic carbon.

3.2 Salinity of the recovered water — recovery efficiency

The development of managed aquifer recharge operations depends on the capacity to store useful quantities of water and to recover useful quantities of good quality water. When water infiltrates or is injected into an aquifer, it mixes with the ambient groundwater, as a result of localised variations in pore-water velocities (dispersion) and solute concentration gradients (diffusion). Mixing may also occur in saline aquifers, due to convection arising from density contrasts. The degree of mixing between the two waters therefore depends on the local hydraulic characteristics of the aquifer, and on the quality of the ambient groundwater.

Restricting mixing is important for the storage of freshwater in brackish or saline aquifers. Where the ambient groundwater is too saline (or contains other constituents at excessive concentrations) for the intended use of recovered water, mixing must be restricted to maximise the volume of recovered water meeting water quality requirements. Mixing is of little significance when the source and ambient waters are both of good quality, unless geochemical processes within the subsurface produce reaction byproducts (eg iron, arsenic) exceeding guidelines.

The primary questions answered within this section are listed below:

- How much of the recharge water can the managed aquifer recharge operator recover?
- What is the extent of mixing between the recharge water and the more saline ambient groundwater?
- What proportion of recovered water will be fit for use?
- What is the distribution of the recharge water within the aquifer, and how far will it migrate?
- Can the recharge water be contained, or will it be drawn by other groundwater users?

3.2.1 Definition of recovery efficiency

Recovery efficiency can be defined as the proportion of recovered water that is of suitable quality for its intended use, expressed as a fraction of the injected or infiltrated volume (Pyne 1995). Thus, a recovery efficiency of 1 indicates an equivalent volume of recovery to recharge, and a value of 0 indicates that none of the recharge water could be pumped at a quality suitable for use.

In practice, recovery is limited by the maximum permissible concentration of the dissolved solids in the recovered water, according to the requirements for its intended use. For example, potable use would imply an upper limit of 500 mg/L total dissolved salts, to meet Australian drinking water standards. Irrigation supplies would be highly dependent on the salt tolerance of any particular land use application. In cases where recovered water is blended with fresher water, a higher permissible concentration may be tolerated, resulting in an increase in recovery efficiency.

There are numerous physical, chemical and microbial species that may be used to 'trace' the recharge water within the subsurface. Many are naturally present in the source water; others can be deliberately added. Some of the more commonly-used tracers of recharge water include:

- total dissolved salts;
- electrical conductivity;
- chloride;
- bromide;

- fluoride;
- temperature;
- deuterium;
- oxygen-18.

Reactive chemicals, and those that contribute to buffering of the system, do not behave conservatively and should not be used.

Numerous factors affect the movement and mixing of recharge water in aquifers, ultimately controlling recovery efficiencies. These variables can be divided into two categories: hydrogeological and management.

Hydrogeological variables are intrinsic to a site, and include:

- aquifer thickness, transmissivity, porosity and heterogeneity (dispersivity and diffusivity);
- regional and local pre-existing hydraulic gradients;
- ambient groundwater quality and density;
- solute concentration of the recharge water and ambient groundwater;
- density difference between the recharge water and ambient groundwater.

Management variables include:

- well design;
- recovery well;
- injection well where applicable;
- total stored volume;
- open interval of the injection and recovery well;
- infiltration and injection rates and volumes;
- recovery rate and volume;
- location of recovery well(s);
- residence time within the aquifer;
- consecutive years or cycles of operation.

Hydrogeological variables cannot be controlled at any given site, but should be considered during site selection. Management variables, however, can be controlled to varying degrees. For example, residence time may be adjusted according the groundwater velocity by adjusting the storage time (well-injection scenario) or the separation distance between recharge basin and pumping well (basin scenario). The role of the unsaturated zone in basin systems may need to be considered, for example, with respect to the potential for mobilisation of salts stored within the soil profile.

3.2.2 Recovery efficiency data for Australian ASR operations

Recovery efficiency data has emerged from several Australian experimental and operational aquifer storage and recovery sites over the past decade. Data for six Australian sites (recalculated from the concentration and volumetric data for a common threshold concentration of 1 500 mg/L) show that:

• recovery efficiency varies significantly (from <0.02 to >1.0) between sites, due to factors discussed above;

• moderate to high recovery efficiency can be achieved in limestone aquifers in which major secondary porosity features are absent;

• improvements in recovery efficiency can occur in successive cycles of aquifer storage and recovery operation, due to salinity buffering by unrecovered recharge water (see Figure 3.1);

• low recovery efficiency is characteristic of highly saline aquifers, due to a combination of the low degree of mixing that can be tolerated density effects that can cause buoyancy stratification of the recharge water.

Similar data for other forms of managed aquifer recharge (eg for surface recharge methods in coastal sand aquifers) are not yet available.

3.2.3 Management of mixing and recovery efficiency

Guiding principles in site selection and operational management to maximise recovery efficiency are drawn from national and international experience, and include the following data:

• avoid highly saline groundwaters;

• reduce salinity of source water;

• select aquifers with low ambient groundwater velocities;

• avoid highly heterogeneous aquifers.

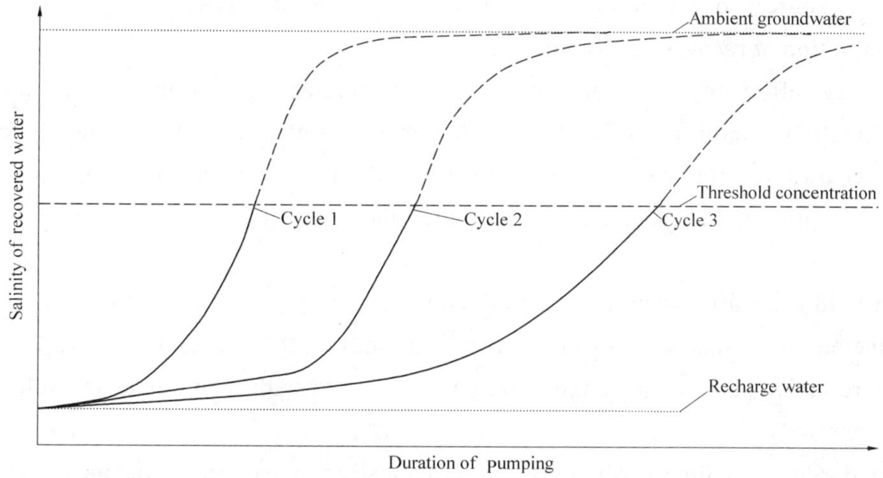

Figure 3.1 Change in salinity of pumped water during the recovery stage, with pumping stopped when the maximum permissible concentration is reached. The unrecovered recharge water forms a salinity buffer, and recovery efficiency improves with subsequent cycles. The dashed curves project the salinity response if pumping was allowed to proceed beyond the threshold concentration

• avoid highly permeable aquifers, particularly when ambient groundwater is highly saline.

- Ensure target aquifer (and unsaturated zone where applicable):
  —is appropriately characterized;
  —has appropriately defined storage intervals;
  —has appropriately designed injection/recovery well(s).
- Ensure adequate volumes of water are stored (performance depends on size).
- Position recovery wells to maximise capture of recharge water.
- Minimise storage time to constrain regional drift of recharge water.

Some of these principles will conflict with other objectives of managed aquifer recharge, requiring some trade-off. For example, to avoid poor recoverability of recharge water, do not use aquifers or subaquifers that are highly permeable; this contrasts with the principles of minimising well clogging outlined previously.

Design scenarios for managed aquifer recharge systems that offer some degree of flexibility to meet water quality objectives. For example, recovery wells for pumping water after surface infiltration can be located so that they largely pump recharge water, minimising the spread of recharge water beyond the storage zone. This is useful when mixing with ambient groundwater is not desirable. Alternatively, wells can be located where they consistently pump a mixture of recharge water and ambient groundwater.

Recovery efficiency may also be improved by blending recovered water with alternative lower salinity sources, or providing post-treatment by reverse osmosis.

3.2.4 Evaluation of recovery efficiency

Recovery efficiency is controlled by many interacting variables that have large uncertainties. It is therefore difficult to predict recovery efficiency in advance of managed aquifer recharge development. Two main methods are available to evaluate recovery efficiency — pilot field trials, and groundwater modelling (using analytical or numerical techniques).

Pilot field trials are essential for establishing operating procedures. However, when they are conducted at a small scale, they often can not predict operational scale recovery efficiency reliably. This is due to the scale dependence of hydrogeological variables such as aquifer dispersivity.

Groundwater modelling predictions can be indicative of recovery efficiency when aquifer hydraulic and transport properties are reasonably well known. In most cases, these predictions would not be expected to preclude the need for field trials. Further information on recovery efficiency pertaining to well injection methods is given by Pyne (1995) and Pavelic *et al.* (2002a).

3.3 Pressure, flow rates, volumes and groundwater levels

3.3.1 Confined and semi confined aquifers

Over-pressurisation of injection wells can rupture an aquitard (a low-permeability geological layer that confines or separates aquifers). This can connect previously unconnected aquifers, leading to adverse changes in piezometric levels and water quality; the consequences of this can outweigh all the benefits of an aquifer storage and recovery operation.

Excessive pressure can also cause failure of poorly completed injection or other wells, allowing water to escape to other aquifers or the ground surface. Pressurising a confined aquifer can also cause existing completed wells in the same aquifer to become artesian. The discharge of water from these may create a nuisance or hazard that needs to be addressed.

On recovery of stored water, the aquifer's piezometric pressures are reduced. This increases energy requirements for pumping of bores that are hydraulically influenced, reducing their yields. In some circumstances, despite the additional storage in the aquifer, existing well owners may be disadvantaged by the aquifer storage and recovery operation.

Excessive lowering of pressures may lead to consolidation of compressible aquitards and other overlying strata, or desaturation of confined aquifers. Either of these situations may cause land to subside. Excessive lowering of pressures can also ultimately trigger failure of an injection well or aquitard, if the aquifer beneath the confining layer is subject to chemical or physical erosion.

Management controls for pressure, flow or volume-related hazards in confined and semi confined aquifers include the following preventive measures, each of which is discussed below:

- selection of injection pressure;
- estimation of the region likely to be made artesian;
- determination of the transfer of fluid pressure and constituents between aquifers;
- restriction of volume, recharge and recovery rates to appropriate periods;
- constrain plastic deformation;
- calculate maximum injection pressure.

3.3.1.1 Injection pressure

The injection pressure should be selected to ensure that it never exceeds the dry overburden pressure on the base of the aquitard. This pressure (P) can be conservatively estimated from $P < 15\ d$ (kPa), where d is the depth in metres from the land surface to the base of aquitard overlying the aquifer, and assuming that the dry weight density exceeds 15 kN/m^3.

This maximum pressure should not be constrained by valves, because inevitably, valve adjustments will be made. An oversized pump could then cause irreparable damage. Normally, the injection pressure will be more tightly constrained by energy efficiency and the need to avoid compaction of the clogging layer around the well perimeter.

In exceptional cases, where more accurate calculations of maximum injection pressure are needed, it may be necessary to:

- measure the bulk density of the aquitard and the material above it (overburden);
- measure the elevation of the piezometric surface in the overburden;
- undertake geotechnical testing to assess the strength of the aquitard;

• calculate bursting pressure by geotechnical evaluation, to account for the elastoplastic properties of the aquitard.

3.3.1.2 Artesian region

The region likely to be made artesian should be estimated using estimated injection rates and aquifer hydraulic properties derived from pumping tests. All wells in the region that penetrate the target aquifer should be identified, permission should be gained from owners, and any well forecast to overflow should be equipped with well-head seals. If this is not viable, it will be necessary to restrict the rate of recharge so that nearest unsealed wells are not artesian.

3.3.1.3 Transfer of fluid pressure and constituents

Checking construction records of influenced wells for completion in multiple aquifers will determine whether transfer of fluid pressure and constituents between aquifers is acceptable, and whether well backfilling may be necessary.

3.3.1.4 Volume, recharge and recovery rates

It is important to restrict volume and rate of net recharge and recovery to periods (eg annual, monthly, daily), that are appropriate to the local conditions and aquifer capacity. Normally, reducing the head in a confined aquifer (so that the aquifer becomes unconfined) is unacceptable, as this may result in consolidation and adverse geochemical changes such as oxidation of reduced iron minerals. Also, the pump should be set above the level of the top of the aquifer.

3.3.1.5 Plastic deformation

If the aquitard is under-consolidated, a quantitative estimate of the amount of plastic deformation likely to occur following the proposed reduction in head during withdrawal should be calculated. Pumping rates should be restricted, to constrain deformation to an acceptable level.

3.3.2 Unconfined aquifers

Recharge of unconfined aquifers increases storage and may protect groundwater-dependent ecosystems in stressed aquifers. However, if the watertable is raised too high, recharge of unconfined aquifers may also have adverse impacts. Examples include:

• waterlogging;

• flooding of basements, below ground cable ducts and depressions;

• reduced traffic access, potholes and increased road maintenance costs;

• effects of anoxia on vegetation;

• increased groundwater ingress to sewers;

• increased differential heave effects on footings, causing consequent structural damage to buildings.

On recovery of stored water, lowering of the watertable may increase pumping costs for other groundwater users and reduce yields of shallow wells. It may also mobilise metals and reduce groundwater discharge to dependent ecosystems at times when this is most needed.

Management controls for pressure, flow or volume-related hazards in unconfined aquifers include the following preventive measures:

- Install overflow diversion to prevent excessive recharge head.
- Prevent use of pumps for injection.
- Avoid selecting infiltration sites that are:
  —on sloping land;
  —underlain by shallow watertables or impervious layers;
  —adjacent to surface depressions or groundwater dependent ecosystem.
- Restrict volume and rate of recharge to periods appropriate to the local conditions (eg annual, monthly, daily).
- Increase drying time in recharge basins or infiltration galleries between infiltration events, to allow groundwater mound to subside.

The height of a groundwater mound beneath a rectangular infiltration basin can be calculated from an equation for any location and time since recharge commenced (Hantush, 1967; Bouwer, 1978, 2002). The result depends on infiltration rate, aquifer transmissivity and porosity, and the size of the basin.

Figure 3.2 shows the mound height beneath the centre of a 100 m×100 m infiltration basin overlying an unconfined aquifer with an initial saturated aquifer thickness of 10 m, hydraulic conductivity of 10 m/d and effective porosity of 0.1. Soil type and source water quality, which affect recharge rates through clogging, therefore also affect mound height. For this example, if the mound height was constrained to 4 m or 8 m, recharge rates would need to be <30 m/a or <80 m/a respectively to comply.

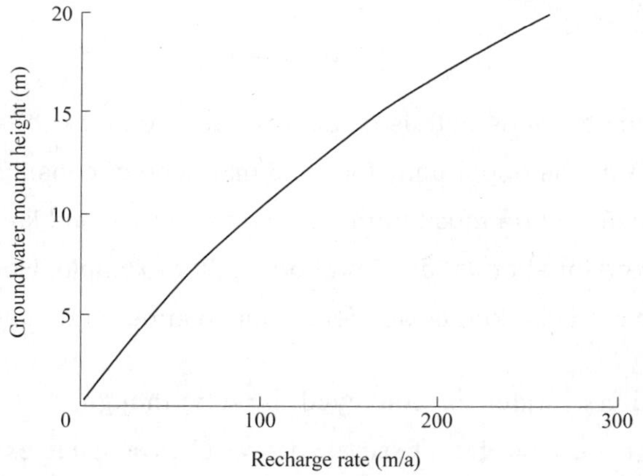

Figure 3.2 Example of calculated groundwater mound height at centre of infiltration basin after 1 year of recharge, for a range of recharge rates

3.4 Contaminant migration in fractured rock and karstic aquifers

In fractured rock aquifers, most water flows through cracks in the rocks, and thus may only interact with a very small percentage of the rock mass. The effective porosity of fractured rock aquifers is much less than for aquifers in which flow is predominantly through interconnected pores throughout the porous medium.

Water is conveyed through connecting fractures, and may zigzag through aquifers in three dimensions. Hence, the water injected into a fractured aquifer will travel further from the aquifer storage and recovery well than the same volume would travel in a porous aquifer (see Figure 3.3). The distribution of recharged water is therefore much more difficult to predict, which has implications for establishing the size of the attenuation zone. If the aquifer is initially brackish, the recovery efficiency will also be diminished. Aquifer pumping tests will normally reveal low effective porosity; but, if results are difficult to interpret, the lowest realistic estimate of effective porosity should be used when estimating the required size of the attenuation zone.

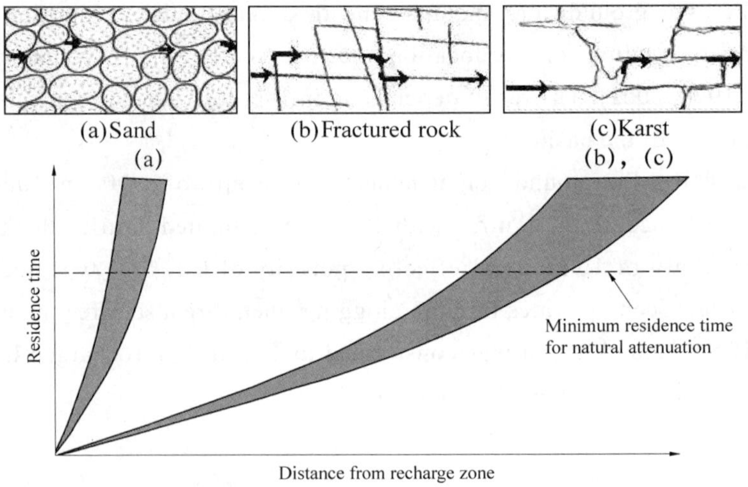

Figure 3.3 Homogeneous porous media (a), fractured rock (b) and karst (c) aquifer characteristics, showing the opportunity for rapid migration of constituents in recharged water. Recharged water can migrate much further in fractured rock and karstic aquifers, within the residence time required for attenuation of pathogens, for example. Hence, the attenuation zone required for fractured and karstic aquifers is much larger and may be difficult to define

In karstic aquifers, water is conveyed largely though dissolution features such as fractures that have been enlarged, fissures, flow tubes and caves. The dimensions of the pathways of least resistance to flow, or preferential flow paths, can be many metres across. Karstic aquifers also commonly allow inter-granular flow; the whole aquifer may transmit water, but most of the flow occurs in the dissolution features. In dual porosity systems such as this, it may take longer to displace the native groundwater from the porous media.

The surface area of the aquifer matrix to which the fastest flowing water is exposed in these preferential flow paths is therefore much smaller than for water flowing through the aquifer matrix. Consequently, opportunities are reduced for the aquifer biofilm to react with constituents of the fastest flowing recharged water. Bulk biodegradation rates may therefore be limited by surface area, and hence slower, than in water flowing through the matrix. Constituents can potentially travel further before biodegrading to levels complying with water quality requirements for ambient groundwater environmental values. The attenuation zone for karstic aquifers may therefore be larger than would be predicted for an aquifer without karstic features.

This situation does not mean that karstic aquifers should be avoided for managed aquifer recharge. However, it does suggest that there will be a greater onus on proponents to demonstrate that managed aquifer recharge operations will not adversely affect other water supplies or connected water bodies. Hydraulic characterisation of the aquifer alone will rarely be sufficient to assess the size of the attenuation zone, but guidance by historical information from nearby managed aquifer recharge operations will be of great value in assessing the likely risk to human and environmental health.

When attenuation zones are relied on for pathogen inactivation or nutrient and contaminant biodegradation, their extent should be no greater than is acceptable to stakeholders and regulators. Fractured and karstic aquifers may require more than one observation well to be confident that concentrations beyond the attenuation zone meet the aquifer's environmental values.

3.5 Aquifer dissolution and well and aquitard stability

Recharge water may react with the aquifer matrix material, resulting in dissolution of minerals or reduction in the aquifer's bulk volume or strength.

Causes of aquifer dissolution may include:

- acidic recharge water dissolving carbonate minerals (eg calcite and dolomite);
- organic carbon in recharge water oxidising to produce acidic conditions within the aquifer; this will also dissolve carbonate minerals;
- aerobic recharge water oxidising sulfide minerals;
- recharge water that is not in equilibrium with mineral phases, leading to mineral dissolution fresh recharge water reacting to dispersed clay minerals in brackish or saline media and changing the physical structure and strength of the porous media;
- cation exchange reactions (especially where recharge water is sodic) dispersing clay minerals and changing physical structure of the porous media.

Aquifer dissolution may increase the effective diameter of a well, consequently increasing yield, and inhibit chronic clogging problems. However, aquifer dissolution has many negative effects, including:

- collapse of uncased wells, due to reduced strength of the surrounding aquifer material;
- production of turbid water or water containing a lot of sand, which can abrade pump

impellors, block irrigation drippers and sprays, and cause nuisance within the distribution systeml;

● eventual undermining of the aquitard in the vicinity of an aquifer storage and recovery well, causing aquitard collapse during recovery, and connecting previously separate aquifers of potentially different water qualities; this has potential for significant enduring environmental damage;

● contribution to metal mobilisation problems;

● adverse effects of taste, odour and safety of recovered water for drinking, particularly following oxidation and reduction of sulfide minerals;

● mobilised clay particles may become trapped further within the aquifer matrix, leading to a chronic decline in specific capacity of injection wells.

3.6 Interactions with other groundwater users including groundwater dependent ecosystems

Operators of managed aquifer recharge projects need to take account of the project's effect on local groundwater users, and of the user's effects on the operation of the project. Other managed aquifer recharge operators also create particular issues, which must be addressed in catchment and groundwater management plans to avoid potential future conflicts. MAR operations that are located close together may result in competition for source water or aquifer storage space.

Nearby recharge operators are likely to be recharging concurrently when there is excess surfacewater, and recovering concurrently when there is a deficit of surface water. Hence, there will be superimposing effects on piezometric heads in the aquifer. In the case of aquifer storage and recovery, for example, each operation will decrease the head gradient of the other during injection, and increase the head against which recovered water needs to be pumped by the other. Each operation therefore increases the costs and energy requirements of the neighbouring operation.

Groundwater-dependent ecosystems (eg springs, streams, wetlands, lakes, phreatophytic vegetation) are potentially affected by changes in groundwater quality and levels, and rates of groundwater discharge exceeding the normal seasonal and interannual range of variation.

MAR may therefore affect ecosystems; for example, by sustaining water in groundwater-dependent wetlands. Depending on ecosystem management objectives, this may have desired and undesired effects. Biodiversity conservation (including migratory birds) is a major consideration for Ramsar wetlands (the Ramsar Convention is an international treaty for the conservation and sustainable use of wetlands), and would need to be evaluated when assessing nearby managed aquifer recharge projects.

4 Water Quality Considerations

Applying the system analysis and management components (Elements 2 ~ 6) of the framework for water quality management to MAR reveals a number of water quality hazards (Table 4.1). Hazards to human health or the environment may originate from:

- the source water for recharge;
- native groundwater;
- aquifer minerals reacting with recharge water;
- byproducts of treatment processes or maintenance practices.

Table 4.1 Summary of key water quality hazards in source water, groundwater and aquifer materials for MAR projects, with examples of specific hazards and preventive methods

Hazard	Origin[a]	Examples	Preventive measures
Pathogens	S, (G)	Viruses	Adequate aquifer residence time
Inorganic chemicals	G, A, S	Arsenic	Control Eh during recharge (avoid mobilisation)
Salinity and sodicity	G, (S)	Salinity	Increase volume of freshwater recharged
Nutrients	S, (G)	Nitrogen	Pretreat water (eg activated sludge)
Organic chemicals	S, (G)	Pesticides	Exclude prone sub-catchments
Turbidity and particulates	S, (G)	Suspended solids	Pretreat water (eg wetland)
Radionuclides	G, A, (S)	Alpha-radiation	Aquifer selection (avoidance)

Note: A=aquifer minerals, G=groundwater, S=source water for recharge;with a Brackets show possible secondary **source**: Eh=a measure of redox potential, the propensity for oxidation and reduction reactions.

Hazardous events such as contaminant migration in fractured rock, or impacts on other groundwater users, are discussed in Chapter 3 Hydraulic Considerations.

These key hazards impact on:

- the aquifer beyond the attenuation zone (and hence other groundwater users and groundwater dependent ecosystems);
- the uses of recovered water;
- situations where the byproducts of MAR water treatments and operations are reused or discharged (see Table 4.2).

Proponents developing a MAR project should engage specialist consultants to identify and quantify health, environmental (eg hydrogeological and geotechnical) and management risks. The next step is to develop preventive measures to mitigate risks and appropriate

Table 4.2 Examples of hazardous byproducts of managed aquifer recharge operations

General hazard	Example of specific hazards
Water treatment byproducts	Any process with reject water (eg reverse osmosis) or byproduct (eg coagulation, filtration, backwash water) may produce water with elevated concentrations of suspended solids, pathogens, inorganic chemicals, nutrients, salinity and organic chemicals
Purge water	Suspended solids, pathogens, metals, nutrients and organics in recharge water may be concentrated in water purged from an aquifer storage and recovery well during maintenance
Basin scrapings	Pathogens, metals, nutrients and organics in recharge water may be concentrated in scrapings produced by infiltration basin maintenance. If they meet quality criteria, scrapings may be reused in agriculture as a component in soil conditioner

Note: The waste management hierarchy should be invoked in the priority order of: avoid, reduce, reuse, recycle, treat and dispose.

monitoring methods to assess each hazard, including critical control points where applicable. Hazard preventive measures may include one or more of the following barriers:

- source water selection;
- recharge control system (eg recharge shut down if the monitored indicator variable is outside critical control limits);
- aquifer selection;
- project location (away from sensitive groundwater-dependent ecosystems or end uses);
- treatment of recharge water to remove hazards or precursor(s) to their formation or occurrence;
- adequate detention time for passive treatment within the aquifer;
- treatment of recovered water before distribution to end uses;
- incident response plans, including feedback from real-time monitoring.

In many MAR operations, multiple barriers may be needed, so that if any one barrier fails, human health and the environment will still be protected. Table 4.3 summarises preventive measures, many of which apply to more than one hazard. Depending on the particular project, other critical control points and preventive measures may be more appropriate. Education and training are important components of implementing and maintaining prevention measures.

4. 1 Pathogens

In confined aquifers, sources of enteric pathogens (ie. intestinal pathogens) are limited to those present in the recharge source water. However, in unconfined aquifers, other sources of such pathogens may exist. Potential water sources for managed aquifer recharge (eg wastewater, greywater, stormwater) can contain a wide range of enteric pathogens that pose a risk to human health.

Table 4.3 Summary of preventive measures and critical control points for MAR

Preventive measures			Description	Critical control point
Exclusion barriers	Preventing entry	Hazard source control	Selection or management of water sources before recharge (eg catchment, recycled water, roof runoff, stormwater)	No
		Intake levels	Exclusion of floating hazards by maintaining intake levels below the water surface	No
		Exclusion of water that does not meet critical limits	Continuous monitoring of indicator variable, to provide feedback to divert water flow or stop recharge when critical limit is exceeded	Possibly: depends on the hazard, associated risk prevented and monitoring system
	Removing hazards	Residence time in the soil or aquifer	Attenuation of all human pathogens and selected organic chemicals	Yes: system needs to provide required residence time. Recovery rate is restricted to ensure adequate time between recharge and recovery
		Travel distance in aquifer	Travel distance (between recharge and recovery) chosen to provide a minimum residence time under the range of operating conditions	No: component of system design
		Treatment processes	Concentration of specific hazards decreased before recharge or on recovery	Possibly: at the point of recharge and/or recovery; depends on the hazard and effect on the specific environmental endpoint
Preventive measures to manage risks in commissioning and validation monitoring (Stage 3)		Operate early warning feedback system	Monitor near-recharge wells for early warning of treatment effectiveness and to allow corrective actions to be implemented	Yes: detection exceeding critical limit trigger corrective action
		Recover recharged water and retreat	Recover contaminated water to prevent exposure	No: however, volume to be recovered would be reduced by use of an early warning system
		Prevent distribution of recovered water to unacceptable end uses	If recovered water does not meet water quality requirements for intended uses, stop recovery (to allow longer residence time), divert to acceptable uses or retreat	Possibly: surrogate parameter correlated with water quality concern may be used
		Post-treatment of recovered water	Treat recovered water to remove identified hazards	Yes: direct or surrogate parameters exceeding critical limit trigger corrective action
		Reduce rate of recharge or recovery	Modify flow rates to increase residence times, reduce pressure gradients across thin aquitards, or reduce MAR-induced flow and level variations	No: subject to validation monitoring

Public health and environmental risks associated with pathogens in relation to managed aquifer recharge are identical to those described in the Australian Guidelines for Water Recycling (NRMMC-EPHC-AHMC, 2006) and shown here in Table 4.4.

Table 4.4 Microorganisms of concern in raw sewage (NRMMC-EPHC-AHMC, 2006)

Pathogen type	Examples	Illness
Bacteria	Salmonella	Gastroenteritis, reactive arthritis
	Campylobacter	Gastroenteritis, Guillain–Barré syndrome
	Pathogenic Escherichia coli	Gastroenteritis, haemolytic uraemic syndrome
	Shigella	Dysentery
	Yersinia	Gastroenteritis, septicaemia
	Vibrio cholerae	Cholera
	Atypical Mycobacteria	Respiratory illness (hypersensitivity pneumonitis)
	Legionella spp	Respiratory illness (pneumonia, Pontiac fever)
	Staphylococcus aureus	Skin, eye, ear infections, septicaemia
	Pseudomonas aeruginosa	Skin, eye, ear infections
	Helicobacter pylori (?)	Peptic ulcers
Virus	Enterovirus	Gastroenteritis, respiratory illness, nervous disorders, myocarditis
	Adenovirus	Gastroenteritis, respiratory illness, eye infections
	Rotavirus	Gastroenteritis
	Norovirus	Gastroenteritis
	Hepatitis A	Infectious hepatitis
	Calicivirus	Gastroenteritis
	Astrovirus	Gastroenteritis
	Coronavirus	Gastroenteritis
Protozoa	Cryptosporidium	Gastroenteritis
	Giardia	Gastroenteritis
	Naegleria fowleri	Amoebic meningitis
	Entamoeba histolytica	Amoebic dysentery
Helminths	Taenia (T. saginata)	Tapeworm (beef measles)
	Ascaris	Roundworm
	Trichuris	Whipworm

4.1.1 Fate and behaviour of pathogens in MARA

Pathogen survival in groundwater is affected by physical, chemical, and biological processes (Toze and Hanna, 2002; Gordon and Toze, 2003). The inactivation of pathogens in aquifers highlights their potential use as robust treatment barriers in the multibarrier approach. Pathogen presence and survival in aquifers is highly variable and is influenced by a variety of factors, including:

- pathogen type;
- recharge water source type;
- temperature;
- redox state and oxygen concentrations;
- activity of indigenous groundwater microorganisms;
- aquifer geochemistry.

Further information on the effects of these parameters on pathogen survival in aquifers is given by Dillon and Toze (2005) and NRC (2008). Although all of these parameters (either independently or collectively) may influence pathogen survival, different pathogens can also vary in their environmental stability (eg between different locations), depending on local groundwater conditions. Enteric viruses are generally accepted to be the most resistant to decay (ie have the highest potential for survival); followed by protozoa and then bacteria.

Little is known about the fate of helminths during MAR. In Australia (except for the tropical northern regions), helminths pose minimal risk. They can be controlled by suitable pretreatment, and are effectively removed from water by simple treatment systems such coagulation, flocculation and sedimentation in stabilisation ponds (Jimenez, 2003). Even if a helminth egg was able to pass through treatment systems, their small size (40 ~ 90 μm) means that simple filtration processes in more consolidated and sand aquifers would be likely to remove them.

Managed aquifer recharge projects relying on aquifer treatment to remove pathogens before recovery for drinking water supplies will always require validation. The efficiency of pathogen removal depends on site-specific conditions; uncertainties can be resolved by monitoring to validate attenuation rates. In terms of log removals, aquifers are conceptually much like other natural and engineered water treatment processes for pathogens (NRMMC-EPHC-AHMC, 2006). Log-removals in aquifers are primarily related to the residence time of the recharge water, the redox state of the aquifer and the temperature; they are typically expressed in terms of the number of days required for a 1-log reduction in pathogen numbers. Table 4.5 shows the ranges of time for 1-log reduction to occur for a selection of pathogens and indicator organisms found in aerobic and anaerobic aquifers.

Table 4.5 Examples of inactivation rates for microorganisms in aerobic and anaerobic aquifers

Aquifer redox status①	1-log reduction times (d)					
	Adenovirus	Rotavirus	Giardia	Cryptosporidium	E. coli	Campylobacter
Aerobic	35 ~ 87	112②	—	22②	1 ~ 3	—
Anaerobic	53 ~ 70	72 ~ 118	—	3 ~ 89	4 ~ 11	—

Note: ①The results given here as examples are taken from MAR schemes where the groundwater temperature is greater than 20 °C. Any aquifer with temperatures less than 20 °C where reliance is placed on pathogen inactivation in the aquifer will need to have a pathogen survival assessment undertaken;

②Results of a single study only;

③Source: modified from Toze *et al.* (2002), Gordon and Toze (2003), Wall *et al.* (2004), and Sidhu *et al* (in press).

The pathogen inactivation rates shown in Table 4.5 are relatively broad, because they are influenced by the site-specific conditions described above. More specific pathogen survival data are given in Appendix 1.

As a rule of thumb, the most reliable assessment of pathogen attenuation at a given site is measurement of pathogen survival in situ using inoculated diffusion chambers. Due to temporal variations in pathogen numbers in source water, detecting no pathogens in groundwater samples sheds little light on the aquifer's actual pathogen inactivation rate. Relevant laboratory-derived attenuation data must be obtained at a temperature and redox status relevant to the aquifer, using chambers inoculated with microorganisms from the aquifer.

Generalised data, such as that given in Table 8, is suitable for precommissioning residual risk assessments. As validation data from MAR sites accumulates, greater precision in estimating pathogen attenuation rates in aquifers is expected.

Verification monitoring involves attempting to detect microbial indicators and targeted enteric viruses in the recovered water. This is done by monthly sampling of observation wells within the attenuation zone and of the recovered water.

Secondary treated effluent does not appear to affect the numbers of indigenous pathogens in an aquifer. Laboratory evaluation of the effect of secondary treated effluent on the abundance of indigenous opportunistic pathogens in aquifer material did not result in any significant difference in comparison with controls (S. Toze, CSIRO Land and Water, pers comm., 2007).

Recent studies (Reed, 2007; Reed *et al.,* 2007) have documented the changes in native microbial populations near the Bolivar (South Australia) reclaimed water ASR trial and the Floreat Park (Western Australia) reclaimed water infiltration galleries. These studies confirmed trends in populations related to proximity to nutrient-rich recharge water and changes in redox status. In distal water, where water quality indicated presence of reclaimed water, microbial community populations, biodiversity and activity were unchanged. Although the evidence is not conclusive, it suggests that MAR is unlikely to stimulate recovery of higher numbers of indigenous opportunistic pathogens (eg Pseudomonas, Aeromonas) than would be recovered by pumping native groundwater.

4.1.2 Management of pathogens via MAR

Preventive measures to reduce the risk of pathogenic hazards and achieve performance targets include:

- source control (eg catchment management for stormwater sources);
- removing pathogens using treatment processes (eg engineered or natural treatment processes to achieve the required log removal rate of the reference pathogens);
- reducing exposure through preventive measures on-site (eg controlling public access during irrigation with recovered water).

These three measures are described in Sections 3.4.1 ~ 3.4.3 and Appendix 3 of the Australian Guidelines for Water Recycling (NRMMC-EPHC-AHMC, 2006).

The concept of tolerable risk is central to the management of enteric pathogens via managed aquifer recharge. These guidelines adopt a tolerable risk of 10^{-6} disability adjusted life years (DALYs) per person per year (NRMMC-EPHC-AHMC, 2006).

Given the potentially large numbers of pathogenic hazards in source waters, three reference pathogens — rotavirus, Cryptosporidium and Campylobacter — have been selected to represent viral, protozoan and bacterial hazards respectively. For a detailed description of DALYs, and the calculation of microbial health-based performance targets for the reference pathogens, refer to Appendix 2 of the Australian Guidelines for Water Recycling (NRMMC-EPHC-AHMC, 2006). The unique feature of MAR is the consideration of the aquifer's effects on pathogens.

4.2 Inorganic chemicals

This section is applicable to the major ions (calcium, magnesium, sodium, potassium, chloride, sulphate, bicarbonate, bromide and fluoride); metals (aluminium, cadmium, chromium, copper, iron, manganese, nickel, lead, strontium and zinc); metalloids (arsenic, boron and silicon); and gases (hydrogen sulfide and methane). Nitrogen and phosphorus are discussed separately in Section 4.4.

Unlike pathogens, there is insufficient information on chemical parameters to support DALYs. Tolerable risk is therefore defined in terms of guideline concentrations (ANZECC and ARMCANZ, 2000; NHMRC-NRMMC, 2004; WHO, 2006; EPHC-NHMRC-NRMMC, 2008a).

The key inorganic hazards resulting from aquifer storage are:

- increased arsenic, iron, manganese, trace species or hydrogen sulfide, producing recovered concentrations in excess of the beneficial use guideline value;
- increased iron in recovered water, which impacts on water supply infrastructure (eg irrigation);
- changes to major ion chemistry alter the sodicity or nutrient balance of the recovered water, affecting its suitability for potential uses (eg irrigation).

When elevated metal concentrations exceeding the beneficial use guideline value occur in backwash water from injection wells, or in the initial water recovered from an aquifer storage and recovery well, care should be taken in the treatment, use and disposal of this waste stream.

4.2.1 Sources and fate of inorganic chemicals in MAR

The chemistry of water stored in an aquifer during MAR is affected by chemical reactions, driven by the aquifer's conditions (eg pH, redox state, minerals, organic matter, microbial activity). Reactions can occur between the source water and the native groundwater, and between the source water and the aquifer material. This can change water quality and aquifer permeability. The key risks related to subsurface reactions are described below.

4.2.1.1 Arsenic increase

Mobilisation of arsenic from the aquifer sediments can occur when pyrite in the storage zone is oxidised, or iron (Ⅲ) oxides are dissolved (see decision tree in Appendix 3). This is a key issue for confined target zones in which reduced minerals are present, despite starting with source water at acceptable arsenic concentrations (Arthur *et al.*, 2003), and may lead to concentrations of arsenic greater than the drinking water guideline value.

4.2.1.2　Iron increase

Release of iron from the sediments in the storage zone occurs mainly when organic matter in the source water reacts with iron (Ⅲ) oxyhydroxides and oxides (ie. goethite, hematite). It can also occur from pyrite oxidation or by changing the pH of the storage zone (see decision tree in Appendix 3). Iron release is generally an aesthetic water quality concern, potentially causing an elevated colour; but, it can also contribute to aquifer clogging. Iron increases can be associated with the release of other hazards such as arsenic and radium.

4.2.1.3　Manganese increase

Dissolution of manganese from manganese oxides and oxyhydroxides in the sediments occurs by reaction with organic matter in the source water, or by changing the pH of the storage zone (Ibison *et al.*, 1994). Like iron, manganese may contribute to the colour of the recovered water.

4.2.1.4　Trace ion increase

The ion species affected include: aluminium, antimony, arsenic, barium, beryllium, cadmium, chromium, cobalt, copper, fluoride, iron, lead, manganese, molybdenum, nickel, vanadium, uranium and zinc. All of these can cause health or environmental concerns. Increases in trace constituents frequently coincide with an increase in iron, manganese or arsenic.

Mechanisms for trace ion release include:

- oxidation of sulfide minerals, such as pyrite, due to presence as trace elements within the mineral (similar to arsenic, although arsenic is generally more mobile than cations);
- iron (Ⅲ) oxide dissolution under changing pH or Eh, as these surfaces often contain adsorbed trace species;
- exchange or displacement from the solid surface by another species (eg cation exchange);
- mineral equilibrium, when water in the storage zone is not in equilibrium with the dominant mineral phases;
- dissolution of mineral phases or accumulated particulates under changing pH or Eh.

4.2.1.5　Hydrogen sulfide increase

Hydrogen sulfide gas is produced when organic matter, introduced to an anoxic storage zone, reacts with dissolved sulphate. In the sequence of microbial-mediated redox reactions, iron mobilisation is likely to precede the production of hydrogen sulfide. Hydrogen sulfide contributes an aesthetic hazard by imparting taste and odour to the recovered water.

4.2.1.6　Changes to major ion composition

Mixing of the source water with saline groundwater, or ion exchange between the source water and the solid phase exchange sites, can significantly affect the contribution of sodium, calcium and magnesium. This is an important consideration if the recovered water is to be used for irrigation, because it can alter the risk of soil sodicity.

4.2.1.7 Excessive dissolution of carbonate minerals

Excessive dissolution of carbonate minerals can:

• lead to injection well and aquitard stability concerns;

• increase production of sand in recovered water;

• extend preferential flow paths, which may reduce the residence time available for hazard attenuation.

Carbonate mineral dissolution may also expose reactive surfaces such as sulfide minerals, and increase the potential for release of metal and metalloid species.

The source water used in a MAR scheme is unlikely to be in equilibrium with the minerals present in the storage zone. As a result, some dissolution of minerals will occur when the source water comes into contact with minerals in the aquifer. The degree of dissolution depends on the solubility of the mineral in the given conditions (eg pH, temperature, pressure, ionic strength, contact time). Mineral equilibrium can also be altered during subsurface storage by other reaction processes. For example, barium can be released by the dissolution of barite ($BaSO_4$) in aquifer sediments, after dissolved sulfate concentrations have been reduced by bacterial sulfate reduction (Zhou and Li, 1992).

Carbonate minerals can be a major influence on the quality of water recovered, because carbonate dissolution is a rapid reaction. In contrast, silicate dissolution is a very slow buffering reaction between pH 6 and 8 (Appelo and Postma, 1999), and has minimal impact on water quality over the timescale of a managed aquifer recharge scheme. Dissolution of carbonate minerals will increase aquifer permeability, and the impact on the stability of injection wells and the aquitard must be considered.

Mineral dissolution can increase the salinity and hardness of the water available for recovery, and also increase minor constituents such as barium or fluoride (fluorite dissolution). Solubility controls may limit the dissolved concentration of some hazards, such as barium (barite solubility), phosphate (apatite solubility) or iron (iron oxyhydroxides or oxides), but mineral precipitation can lead to clogging concerns. The tendency for mineral dissolution or precipitation can be examined through the saturation index of a solution.

The redox state within the storage zone alters the inorganic chemistry of the recovered water. Redox reactions in MAR will often be induced by addition of source water that contains oxygen to an anoxic aquifer, or by addition of organic matter to an aquifer. Redox zones, and the resulting water quality, can vary spatially and temporally during a managed aquifer recharge operation. A highly reactive zone often develops near the point of injection or infiltration, resulting in water quality that differs from the bulk of the stored water. For example, if source water high in organic matter is used, the reactive zone can become anaerobic. This can lead to dissolution of iron (either present in the aquifer sediments or accumulated around the injection well from filtration of particulate matter in the source water) or in situ generation of hydrogen sulfide.

Changes in redox conditions and water quality can also occur:

- over time under different flow rates (eg injection or infiltration versus storage);
- where degree of saturation changes (eg wetting and drying cycles in soil aquifer treatment);
- in flow reversal during recovery in an aquifer storage and recovery operation.

Sorption to clay minerals, organic matter or iron oxide surfaces can act as an attenuation mechanism for trace metals and metalloids. However, sorption is not permanent. It can be reversed by preferential sorption of another species, or by pH–Eh–dependent changes in the surface properties that alter the number of available sorption sites. In addition, the sorption capacity of an aquifer may be limited, delaying breakthrough of the hazard to a downstream monitoring or recovery location.

Mixing is an important influence on the quality of recovered water, if the native groundwater is brackish or contains hazards that exceed target values for the specific beneficial use. Mixing of two waters can produce a solution that is more aggressive toward the aquifer minerals (Runnels, 1969). If an inorganic constituent in the groundwater exceeds water quality targets, dilution may not be sufficient to ensure water quality targets are met if additional processes, such as mineral dissolution, release that species from the aquifer sediments. The effect of mixing should be considered in relation to salinity targets or operational constraints.

Metals in source water are largely in particulate form, and thus will accumulate in the subsurface, close to the point of entry (ie. well face or basin floor). Accumulated metals may be removed permanently from the system by operational maintenance, such as well redevelopment or basin scraping. Some in situ dissolution may occur, producing a localised increase in soluble metal concentration. This is most likely to affect the first water recovered from an aquifer storage and recovery well.

4.2.2 Management of inorganic chemicals

Source control may include limiting the contribution from hazardous activities (eg catchment management, trade waste discharge agreements), and diversion of flow outside water quality criteria (eg pH, conductivity).

Pretreatment measures include:

- source water treatments (eg filtration, coagulation, flocculation);
- pH adjustment (eg prevention of manganese release from sediments) (Ibison *et al.*, 1994);
- redox control to limit reaction within the aquifer (eg limiting organic carbon in source water, deoxygenating to prevent arsenic release).

Residual risk assessment management of inorganic chemicals within the aquifer requires validation of thc conceptual understanding of geochemical processes and of the preventive measures necessary to manage the human health and environmental risks. Validation monitoring (Stage 3) is necessary because the presence of trace amounts of some minerals can cause problems. Drill core and groundwater samples collected in Stage 2 cannot be relied on to detect all minerals present in the aquifer that recharged water will come into contact with.

Upon recovery and (where necessary) for high-value use, the major ions contributing to salinity can be removed. This can be achieved by post-recovery treatments such as aeration and filtration, or other techniques (individually or in combination) including oxidation, precipitation, coagulation, sorption, ion exchange, lime softening and filtration (NHMRC-NRMMC 2004).

4.3 Salinity and sodicity

The public health and environmental risks associated with salinity and sodicity (the abundance of sodium relative to calcium and magnesium) in relation to MAR include:

- salinity values exceeding the beneficial use value for total dissolved salts or sodium content;
- osmotic effects on plant health and yields, due to irrigation with saline water;
- rising watertables, due to leaching requirements to remove excessive salinity;
- sodicity-related decline in structure of agricultural soils;
- salinity effects on infrastructure and other assets (eg excessive corrosion or scaling in pipes, fittings and appliances; salt damp in stone and masonry structures).

The mixing of recharge water and ambient groundwater in MAR will cause the salinity of recovered water to differ from that of the recharge water. In general, the salinity of ambient groundwater within aquifers targeted for managed aquifer recharge should be similar to or higher than the source water (in keeping with the principles outlined in the Groundwater Protection Guidelines, ANZECC-ARMCANZ, 1995). Therefore, native groundwater will represent an additional source of salinity (and sodicity) in recovered water. The environmental risks of salinity and sodicity and their effects on soil structure and agricultural production are discussed in the Australian Guidelines for Water Recycling (NRMMC-EPHC-AHMC, 2006).

4.3.1 Sources of salt in MAR

All source waters for MAR contain natural salinity levels, derived from inorganic salts, minor amounts of dissolved organic matter and small colloidal material. The inorganic constituents of source waters may be characterised by measures such as conductivity, total dissolved salts and sodicity. Typically, the salinity of roof runoff is lower than stormwater runoff. Stormwater runoff, in turn, has lower salinity than water recycled from sewage effluent. This is because the enrichment or addition of salts from natural or anthropogenic processes in water transported through a rural or urban catchment increases its salinity levels.

Salinity levels in groundwater range from fresh to highly saline. Infiltration of shallow saline groundwater into leaky sewers can substantially increase the salinity and sodicity of sewage effluent, rendering it unfit for recycling via managed aquifer recharge unless it is treated or blended.

4.3.2 Management of salinity and sodicity

Management controls include preventive measures such as:

- catchment water quality management and source control (to minimise salt export and remove or mitigate point sources of salinity, where viable);
- source water selection;
- site selection to target aquifers that minimise risk;
- pretreatment or post-treatment (desalination);
- shandying of recovered water with alternative, lower salinity sources (shandying is the addition of one water source to another).

4.4 Nutrients: nitrogen, phosphorus and organic carbon

This section discusses nitrogen and phosphorus (recognised as environmental hazards for water recycling, NRMMC-EPHC-AHMC, 2006) and organic carbon(an important nutrient in relation to microbial processes in the subsurface).

Nitrogen and phosphorus are identified as key environmental hazards due to their potential for causing nutrient imbalance in irrigation water, soil eutrophication and toxic effects on terrestrial biota (NRMMC-EPHC-AHMC, 2006). While subsurface storage is likely to reduce nutrient concentrations, the overall nutrient balance of the recovered water still needs to be considered in relation to its beneficial use. Nutrients (predominantly organic matter) in the source water will stimulate microbial activity in the subsurface. In turn, this alters the concentration of inorganic and organic chemicals in the water and affects aquifer permeability.

4.4.1 Sources and fate of nutrients in MAR

The level and variability of nutrient loads in source waters is largely affected by pretreatment measures. Recycled water potentially contains high nutrient loads that may vary with seasonal effects on microbial treatment processes. Nutrient concentrations in stormwater are generally likely to be lower than in recycled water, but will vary with catchment type (eg industrial areas).

Removing organic carbon and nitrogen is a passive water quality treatment provided by MAR operations (Dillon and Toze, 2005). Organic matter can be removed by biodegradation, microbial assimilation, filtration, sorption or precipitation. Biodegradation occurs through redox processes that influence the mobility of inorganic chemicals and the fate of organic chemicals. Colloidal organic matter can also facilitate the transport of other chemical hazards. The amount of organic matter removed by biodegradation depends on its character; the easily biodegradable portion can be removed within days of introduction to the subsurface, while less reactive material may degrade over a longer time (Fox *et al.*, 2001). Reactive organic matter present in the sediments may also be degraded. Microbial assimilation occurs when the nutrient rich source water is introduced to the subsurface, leading to the development of a biofilm near the point of entry (ie. well face or basin floor). The biofilm forms a reactive zone that can have distinct redox chemistry from the rest of the storage zone, and is largely responsible for subsurface water quality treatment. Filtration of particulate organic carbon provides an additional energy source to sustain microbial activity, and can produce soluble degradation byproducts.

Nitrogen can exist in various forms in source waters, including organic nitrogen (predominantly proteins), ammonium, nitrate, nitrite and gaseous nitrogen. The dominant nitrogen species in recycled water are organic nitrogen, ammonium and nitrate. The fate of nitrogen depends on its form and the redox conditions encountered. Under aerobic conditions, nitrification will convert ammonium to nitrate; under anaerobic conditions, ammonium can be adsorbed to mineral surfaces by ion exchange until the exchange capacity is exceeded. Nitrate can be removed by reduction to nitrogen gas (denitrification) or ammonium. In the unsaturated zone, some partitioning of ammonium to gaseous ammonia and loss through volatilisation at the air-water interface may occur.

4.4.2 Management of nutrients

Source control may include limiting the contribution from hazardous activities, and diverting flow when water quality indicators (eg colour and turbidity) exceed pretreatment or discharge criteria.

Pretreatment measures include:

- inline filtration on source water delivery infrastructure (for particulate organic carbon);
- biofiltration;
- passive treatment through wetlands.

The effectiveness of natural treatments systems (such as wetlands for nutrient removal) depends on their maintenance. Monitoring is necessary to assist with managing wetland treatment systems.

Removal of subsurface organic carbon and nitrogen relies on treatment through redox processes (eg Soil Aquifer Treatment; Fox *et al.*, 2001; Amy and Drewes, 2007). Validation would need to be supported by evidence of declining concentrations and physiochemical conditions. Phosphorus removal would also need evidence of declining concentrations, supported by mineralogy (iron, aluminium oxides) or mineral saturation calculations. Organic carbon can be recovered from recovered water by granular activated carbon and membrane filtration, if necessary, for high-valued use (NHMRC-NRMMC, 2004). Biological clogging of recharge wells and infiltration basins and galleries should also be considered.

4.5 Organic chemicals

This section discusses trace organic compounds (often referred to as micropollutants) including:

- pesticides;
- hydrocarbons;
- polycyclic aromatic hydrocarbons (PAHs);
- emerging chemicals of concern;
- endocrine disrupting chemicals;
- personal care products;
- pharmaceuticals;
- flame retardants.

Trace organic compounds are predominantly anthropogenic in origin (eg PAHs are a combustion product of carbon fuels); however, some may be naturally occurring (eg algal toxins).

Organic chemicals can pose health risks (NHMRC-NRMMC, 2004) and environmental risks (NRMMC-EPHC-AHMC, 2006). There are numerous emerging chemicals—for example, endocrine disrupting chemicals, pharmaceuticals and personal-care products, and some disinfection byproducts(eg N-nitrosodimethylamine, NDMA, NRMMC-EPHC-AHMC, 2006). If an established water quality guideline value does not exist for a specific chemical, EPHC-NHMRC-NRMMC(2008a) provides methods for determining guideline values for any chemical with respect to drinking water uses; it also provides guidance on dealing with mixtures of chemicals.

Environmental toxicity testing may be required to provide additional information on the impacts of MAR projects on sensitive environments. Guidance on environmental toxicity testing is provided in Chapter 3 of Aquatic ecosystems of the Australian and New Zealand Guidelines for Fresh and Marine Water Quality (ANZECC and ARMCANZ, 2000a).

4.5.1 Sources and fate of organic chemicals in MAR

Determining the presence of organic chemicals in source waters, and carrying out the associated risk assessment can be difficult, due to intermittent loadings and analytical detection capabilities for these substances. However, the origin of the source water should allow the likelihood and nature of organic chemical presence to be estimated. The effort taken to characterise organic constituents of source water must be proportionate to the risk posed to public health and the environment.

In general, subsurface storage provides a treatment step for organic chemicals (Dillon and Toze, 2005). However, formation and attenuation of trihalomethanes (a group of disinfection byproducts) has been reported during storage(Pavelic *et al.*, 2005, 2006b). The potential for disinfection byproduct formation can be lowered by reducing the amount of organic matter in the source water, or altering the disinfection regime to reduce residual chlorinc (Jimenez, 2003).

Subsurface removal of organic chemicals can occur through volatilisation (in thc unsaturated zone) and biodegradation (in the unsaturated and saturated zones). Degradation rates vary with pH, temperature, redox state and the presence of cosubstrates such as dissolved organic carbon. Sorption will also retard organic chemical movement, but the sorption removal capacity may be limited. However, sorption does provide additional residence time for degradation to occur. No allowance should be made for attenuation in the aquifer due to sorption alone for chemicals that do not degrade under the redox conditions and temperature relevant to the aquifer's storage zone.

The availability of environmental fate data for organic chemicals varies considerably. The fate of hazards such as benzene, for example, is reasonably well documented (Howard, 1991); but there may be little information on emerging chemicals of concern. It is critical to ensure that existing environmental fate data was determined in a similar physiochemical environment to that expected in the MAR scheme under consideration. In the absence of relevant field data, laboratory studies can predict the fate of the hazard under the expected conditions (Oliver *et al.*, 1996; Ying *et al.*, 2003).

If the storage zone is an unconfined aquifer, organic chemicals may be present from point sources (eg industrial activities) and diffuse sources (eg pesticide use). Mixing between the source and native groundwater may therefore affect the recovered water quality.

4.5.2 Management of organic chemicals

Source control may include limiting the contribution from hazardous activities to trade waste or stormwater discharge, and improving hazard management for high risk activities in order to reduce source concentrations or prevent against shock loadings from spills.

Pretreatment and post-treatment measures include biofiltration, passive treatment through wetlands and advanced tertiary treatment, such as membrane filtration.

Reliance on attenuation of organic chemicals in the aquifer requires validation of the reduced concentration with time and distance, supported by details of the residence time in the aquifer and its physicochemical and redox conditions.

Indicators may be selected for monitoring to focus effort on species that will give the most sensitive indicator of effective system performance (Drewes *et al.*, 2008).

4.6 Turbidity and particulates

The public health and environmental risks associated with turbidity in relation to managed aquifer recharge include:

- Recovered water having turbidity:
  - —in excess of drinking water guidelines (where drinking is an intended end use);
  - —which if not removed, can impact on pumps and irrigation infrastructure.

- Reduced disinfection performance, leading to increased risk from microbial pathogens.
- Increased risk of transporting a range of contaminants that can sorb to particles:
  —heavy metals;
  —phosphorus;
  —various organics;
  —microbial pathogens.
- Discharge of backwash water during redevelopment (backwashing) of injection wells, impacting on the stormwater catchment downstream.

4.6.1 Sources of turbidity in MAR

All source waters for managed aquifer recharge contain natural levels of particulates—measured as turbidity or suspended solids—derived from inorganic silt, clay-sized particulates and organic matter. Stormwater runoff usually contains highly variable turbidity levels, as a result of factors related to climate, catchment geomorphology, and land use and management. Secondary or tertiary treated sewage effluent typically contains lower concentrations of particulates, and a higher organic content, than stormwater. Roof runoff is typically low in particulate matter, but can be high due to deposition of vegetation debris or poor management. Groundwater turbidity levels are generally low, but can be high in wells that are inappropriately designed or inadequately developed.

Managed aquifer recharge practices can generate particulate hazards as a result of mineral dissolution and particle remobilisation within the soil or aquifer, and through the standard practice of backwashing injection wells to maintain recharge rates.

4.6.2 Management of turbidity

Turbidity management controls include preventive measures such as:

- Source selection.
- Catchment water quality management and source control to:
  —minimise particulate export;
  —remove or mitigate point sources of turbidity (where viable).
- Pretreatment or post-treatment before recharge through:
  —settling tanks;
  —wetlands;
  —coagulation;
  —filtration.

4.7 Radionuclides

Radioactive materials (eg uranium, thorium, potassium-40) occur naturally in the environment, and risk of human exposure to radiation is predominantly from natural sources. Additional exposure can occur through anthropogenic activities such as medical (radiopharmaceuticals) and industrial use of radioactive materials.

The main radionuclide concern is recovery of water posing a risk to human health by ingestion of drinking water or foods via crop irrigation, stock watering, or food chain accumulation (radium and radon), or inhalation of gas released from the water supply (radon).

Radioactivity is measured in becquerel (Bq), where 1 Bq= 1 disintegration per second. Health considerations are based on the effective dose of radiation, measured in sievert (Sv), which takes into account the equivalent dose received by all tissues or organs, weighted to account for their different sensitivities to radiation. The acceptable radiation dose via the ingestion of water should be <1 mSv/a (NHMRC-NRMMC, 2004). Dose estimates based on the dosage per unit intake of individual radionuclides can be calculated using Table 7.1 and Section 7.6 from the Australian Drinking Water Guidelines (NHMRC-NRMMC, 2004).

4.7.1 Sources of radionuclides in MAR

Recycled water or stormwater may contain radionuclides if they receive water from medical and industrial uses. Groundwater may contain naturally occurring radium and radon isotopes (radium-226, radium-228 and radon-222). Mining activities may also concentrate naturally-occurring radionuclides (eg processing mineral sands, producing phosphate fertiliser).

The major source of radionuclides in MAR will usually be from the interaction of stored water with the aquifer matrix during aquifer storage. Native groundwater radioactivity is a useful indicator of the minimum level of radiation in the recovered water. Natural concentrations of radionuclides vary considerably, and depend on the properties of the aquifer, which are (Dillon and Toze, 2005):

- geology;
- porosity;
- grain size;
- redox state;
- major ion chemistry.

In general, high radionuclide concentrations are found in granitic fractured rock (crystalline) aquifers and near rich organic coal deposits (Herczeg and Dighton, 1998). Leaching of uranium from carbonate aquifers has also been reported (Williams *et al.*, 2002).

Radon concentrations in recovered water and the native groundwater before managed aquifer recharge will be similar, because equilibrium between radon in the aquifer material and the source water is reached in less than three weeks (Dillon and Toze, 2005).

Iron or manganese oxyhydroxides can adsorb substantial amounts of radium. Thus radium concentrations can increase through a MAR scheme if oxidation of organic matter leads to dissolution of these iron or manganese oxyhydroxide surfaces. Radium concentrations can also increase through the dissolution of radium-bearing minerals such as phosphates (Dillon and Toze, 2005).

4.7.2 Management of radionuclides

The radioactivity of native groundwater in the storage zone can be screened by measuring gross alpha and beta activity (excluding potassium-40), followed by analysis of individual radionuclides, if the gross alpha or beta exceeds the target value (NHMRC-NRMMC, 2004).

If the target aquifer is a potential source of radium, geochemical modelling is warranted to evaluate the potential for additional release through mineral dissolution or oxidation of organic rich deposits. Modelling would also define acceptable values for pH, oxygen, nitrate and organic carbon within the source water, to minimise the potential for geochemical reactions that could release radium during aquifer storage.

Potential pretreatment and post-treatment includes aeration for radon-222, and lime softening, ion exchange or reverse osmosis for radium-226 and radium-228(NHMRC-NRMMC, 2004).

Appendix

Appendix 1 Examples of Methods Available for Investigations and Validation Monitoring, as part of stage 3 Commissioning and Trials

Attached table 1 provides examples of the tools available for conducting a Stage 3 assessment, including validation monitoring. Examples are given on a hazard-specific basis. A basic description and reference to sources of information or case studies is provided for each method.

Attached table 1 Examples of the tools available for a Stage 3 assessment

Hazard	Methods of analysis	Information provided	References
Pathogens	Literature	Indicative source concentrations and decay rates for specific pathogens or indicator species	Appendix 2 and NRMMC-EPHC-AHMC (2006)
	In situ or laboratory chamber decay studies	Site-specific decay rates	Pavelic *et al,* 1998
	Sampling and analysis of source water and native groundwater①	Source concentration, physicochemical characteristics of ambient groundwater	Toze and Hanna (2002) Toze *et al.* (2002)
	Tracer testing, aquifer pump testing, EM flow metering①	Groundwater flow velocity, travel times	Pavelic *et al.* (2006a)
Inorganics	Literature	Indicative source concentrations	Appelo and Postma (1999)
	Decision trees	Qualitative view of processes	Appendix 3
	Sampling and analysis of source water and native groundwater①	Source concentration, physicochemical characteristics of ambient groundwater	Herczeg *et al.* (2004)
	Sampling and analysis of soil and aquifer	Presence and quantity of reactive phases	Bekele *et al.* (2007)
	Laboratory-scale column and batch tests	Kinetics of processes	Bekele *et al.* (2007)
	Geochemical modelling (PHREEQC②, PHT3D③)	Quantitative understanding of processes	Greskowiak *et al.* (2005)

Attached table 1 Continueus

Hazard	Methods of analysis	Information provided	References
Salinity and sodicit y	Literature	Indicative source concentrations	NRMMC-EPHC-AHMC (2006)
	Sampling and analysis of source water and native groundwater①	Source concentration, physicochemical characteristics of ambient groundwater	Herczeg *et al*. (2004)
	Sampling and analysis of soil and aquifer	Presence and quantity of reactive clays and salt loads	Bekele *et al*. (2007)
	Laboratory-scale column and batch tests	Assistance with operational design	Bekele *et al*. (2007), Cavé (2000)
	Geochemical modelling Solute transport modelling	Degree of mixing in subsurface	Pavelic *et al*. (2006a)
Nutrients	Literature	Indicative source concentrations	NRMMC-EPHC-AHMC (2006)
	Sampling and analysis of source water and native groundwater①	Source concentration, physicochemical characteristics of ambient groundwater	Vanderzalm *et al*. (2006)
	Sampling and analysis of soil and aquifer	Presence and quantity of reactive phases	Vanderzalm *et al*. (2006)
	Laboratory-scale column and batch tests	Fate and transport, assistance with operational design	Fox (2002)
Organic chemicals	Literature	Indicative source concentrations andenvironmental fate data	NRMMC-EPHC-AHMC (2006)
	Sampling and analysis of source water and native groundwater	Source concentration, physicochemical characteristics of ambient groundwater	Pavelic *et al*. (2005)
	Sampling and analysis of soil and aquifer	Presence and quantity of reactive phases	Pavelic *et al*. (2005)
	Passive samplers	Integrated source concentrations at low levels of detection	Komarova *et al*. (2006)
	Laboratory-scale column and batch tests	Fate and transport, assistance with operational design	Ying *et al*. (2003)
	Geochemical modelling (PHREEQC②, PHT3D③)	Quantitative understanding of processes	Greskowiak *et al*. (2006); Miller *et al*. (2002)
	Tracer testing, aquifer pump testing, EM flow metering	Groundwater flow velocity, travel times	Freeze and Cherry (1979); Clark *et al*. (2005)

Attached table 1 Continueus

Hazard	Methods of analysis	Information provided	References
Turbidity and particulates	Literature	Indicative source concentrations	NRMMC-EPHC-AHMC (2006)
	Sampling and analysis of source water and native groundwater①	Source concentration, physicochemical characteristics of ambient groundwater	Pavelic *et al.* (2006b)
	Sampling and analysis of soil and aquifer	Presence and quantity of reactive phases	Rinck-Pfeiffer *et al.* (2000)
	Laboratory-scale column and batch tests	Fate and transport, assistance with operational design	Rinck-Pfeiffer *et al.* (2000)
Radionuclides	Literature	Indicative concentrations	Cook and Herczeg (2000)
	Sampling and analysis of source water and native groundwater①	Source concentration, physicochemical characteristics of ambient groundwater	Popit *et al.* (2004)
	Sampling and analysis of soil and aquifer	Reactivity of sediments	Gundersen and Wanty (1993); Herczeg and Dighton (1998)
Hydraulics: pressure, groundwater levels, dissolution, aquitard stability	Literature	Initial parameter estimates	Freeze and Cherry(1979)
	Monitoring of Ground-water levels	Hydrostatic conditions	Freeze and Cherry (1979)
	Aquifer pump testing①	Local aquifer hydraulic properties	Freeze and Cherry (1979)
	Geophysical well logging	Aquifer confinement, well integrity	Pavelic *et al.* (2001)
	Groundwater modelling	Scenario modelling of piezometric fluctuations	Dillon *et al.* (1999)
	Geotechnical evaluation	Geotechnical properties of aquitard	Jones *et al.* (2005)
	Geochemical evaluation and modelling	Effect of mineral dissolution on well stability	Le Gal La Salle *et al.* (2005)
Groundwater-dependent ecosystems	Literature	Initial parameter estimates	Freeze and Cherry (1979)
	Monitoring of groundwater levels	Hydrostatic conditions	Freeze and Cherry (1979)
	Sampling and analysis of source water, native ground-water and receiving surface water①	Source concentration, physicochemical characteristics of ambient groundwater	Fies *et al.* (2002)
	Hydraulic and solute transport modelling	Scenario modeling of piezometric fluctuations and impacts on surface water	Fies *et al.* (2002), Escalante *et al.* (2005)
	Ecotoxicity testing	Environmental impacts	ANZECC-ARMCANZ (2000a)

Note: EM = electromagnetic.

① drilling of new wells may be required;

② a computer program designed to perform a wide variety of low-temperature aqueous geochemical calculations;

③ a reactive multicomponent transport model for saturated porous media.

Appendix 2 Inactivation Rates for Pathogens in Aquifers

Attached table 2 shows attenuation rates (days for 1-log removal) of various enteric microorganisms in groundwater from in situ diffusion chamber studies and laboratory experiment studies. These rates are intended for use in calculating pathogen attenuation in the subsurface for various MAR types and configurations. These results should be used as an indication of the relative persistence of different enteric microorganisms in groundwater, and thus used only as a guide. All of the results were obtained from studies where active groundwater microorganisms were present. Attached table 3 gives a very brief description of the different enteric microorganisms that have been used in the studies. If residual risks of pathogens are unacceptable, then disinfection could be considered (subject to trihalomethane [THM] risks), or improved estimates of attenuation rates could be determined using in situ chambers suspended in wells (Pavelic *et al.*, 1998).

Attached table 2 1 log10 reduction times (T90) of enteric microorganisms

Microorganism	Aerobic			Anaerobic	
	Temp (°C)	T90 (d)	Reference	T90 (d)	Reference
Coxsackievirus[a]	15	10	Gordon and Toze,2003[b]	—	—
Coxsackievirus[c]	19~21	20	Jansons,*et al.*,1989[d]	—	—
Coxsackievirus[a]	27	8	Toze and Hanna,2002[d]	—	—
Coxsackievirus[a]	28	10	Gordon and Toze,2003[b]	—	—
Coxsackievirus[a]	28	7~9	Toze,*et al.*,2002[d]	144	Gordon and Toze,2003[b]
MS2[a]	12	77	Collins,*et al.*, 2006[be]	—	—
MS2[a]	12	6~33	Yates,*et al.*,1985[bf]	—	—
MS2[a]	13	9~13	Yates,*et al.*,1985[bf]	—	—
MS2[a]	15	1	Gordon and Toze,2003[b]	—	—
MS2[a]	17	13	Yates,*et al.*,1985[bf]	—	—
MS2[a]	18	12	Yates,*et al.*,1985[bf]	—	—
MS2[c]	22	6	Toze,*et al.*,2002[d]	—	—
MS2[a]	23	<1	Blanc and Nasser,1996[be]	—	—
MS2[a]	23	2~6	Yahya,*et al.*,1996[bf]	—	—
MS2[a]	23	2~5	Yahya,*et al.*,1996[bf]	—	—
MS2[a]	27	5	Toze and Hanna,2002[d]	—	—
MS2[a]	27	<1	Alvarez,*et al.*,2000[bf]	—	—
MS2[a]	28	3	Gordon and Toze,2003[b]	8	Gordon and Toze,2003[b]
MS2[a]	28	8~9	Toze,*et al*,.2002[d]	—	—
Poliovirus[a]	15	5	Gordon and Toze,2003[b]	—	—
Poliovirus[a]	28	1	Gordon and Toze,2003[b]	77	Gordon and Toze,2003[b]

Attached table 2 continuous

Aerobic				Anaerobic	
Escherichia coli[a]	15	1	Gordon and Toze,2003[b]	—	—
E. coli[c]	22	3	Toze *et al.*,2002[d]	—	—
E. coli[a]	27	1	Toze and Hanna,2002[d]	—	—
E. coli[a]	28	1	Gordon and Toze,2003[b]	6	Gordon and Toze,2003[b]
S. typhimurium[c]	22	7	Toze *et al.*,2002[d]	—	—
S. typhimurium[a]	27	1	Toze and Hanna,2002[d]	—	—

Note: a—Survival rates determined using laboratory studies.

b—Studies on pathogen decay in groundwater not necessarily relating to specific managed aquifer recharge schemes.

c—Survival rates determined using in situ studies.

d—Studies undertaken as part of specific managed aquifer recharge schemes.

e—Redox conditions were not specified in the journal paper and not enough information was provided on the methodology to subsequently determine redox conditions.

f—Redox conditions were not specified in the journal paper but suspected to be aerobic due to the experimental methods used.

Attached table 3 Brief descriptions of enteric microorganisms

Enteric microorganism	Details
Coxsackievirus	An enteric virus closely related to Poliovirus and a common pathogen in most human communities, particularly in the young Causes a range of diseases including hand, foot and mouth disease Has been noted to be more resistant than poliovirus under most conditions An obligate intercellular pathogen
Escherichia coli	An enteric bacterium commonly used as an indicator of faecal contamination and for treatment effectiveness The toxigenic form can be a pathogen in its own right Metabolically active in water
MS2	A virus that only infects certain types of E. coli cells Non-pathogenic to humans or animals Often used as a surrogate for enteric viruses, due to its similar morphology and great ease of detection and study

Attached table 3 continuous

Poliovirus	An enteric virus well documented for causing poliomyelitis (polio) Not a serious health hazard in Australia, due to broad vaccination of the community; but the vaccine strain is commonly detected in sewage effluent One of the enteric viruses least resistant to environmental conditions An obligate intercellular pathogen
Salmonella typhimurium	A bacterial pathogen causing gastroenteritis Metabolically active in water

Appendix 3 Decision Trees to Identify Potential Arsenic and Iiron Release in MAR

Understanding the potential for hazard release through geochemical processes during MAR requires knowledge of the:

- source water quality;
- native groundwater quality;
- composition of the aquifer matrix;
- mineralogy;
- major and minor elements;
- exchange potential;
- organic carbon content.

Attached figure 1 and Attached figure 2 provide decision trees to support the discussion in Section 4 to identify the potential for release of arsenic or iron in a MAR scheme. The spatial distribution of arsenic within the sediments of the storage phase may vary considerably, and thus the potential for arsenic release can be difficult to predict. The decision trees do not substitute for geochemical modelling, but may be used to understand the important geochemical processes.

Arsenic release (see Attached figure 2) may also indicate release of other hazards such as: cobalt, chromium, copper, nickel and zinc. It is recommended that groundwater arsenic concentrations, which are used as a surrogate to indicate arsenic in the sediments, be reported to a 1 μg/L detection limit. A concentration reported to a higher detection limit (ie.<5 μg/L) does not preclude the potential for sufficient arsenic release to exceed guideline values.

The decision tree for iron (see Attached figure 2) can be used to consider the potential for iron release and the subsequent recovery of water elevated in iron, or aquifer clogging due to the precipitation of iron oxides in the vicinity of the recharge point. As manganese exhibits similar behaviour to iron, the iron decision tree could be modified to examine the potential manganese release if manganese is a major constituent of the aquifer.

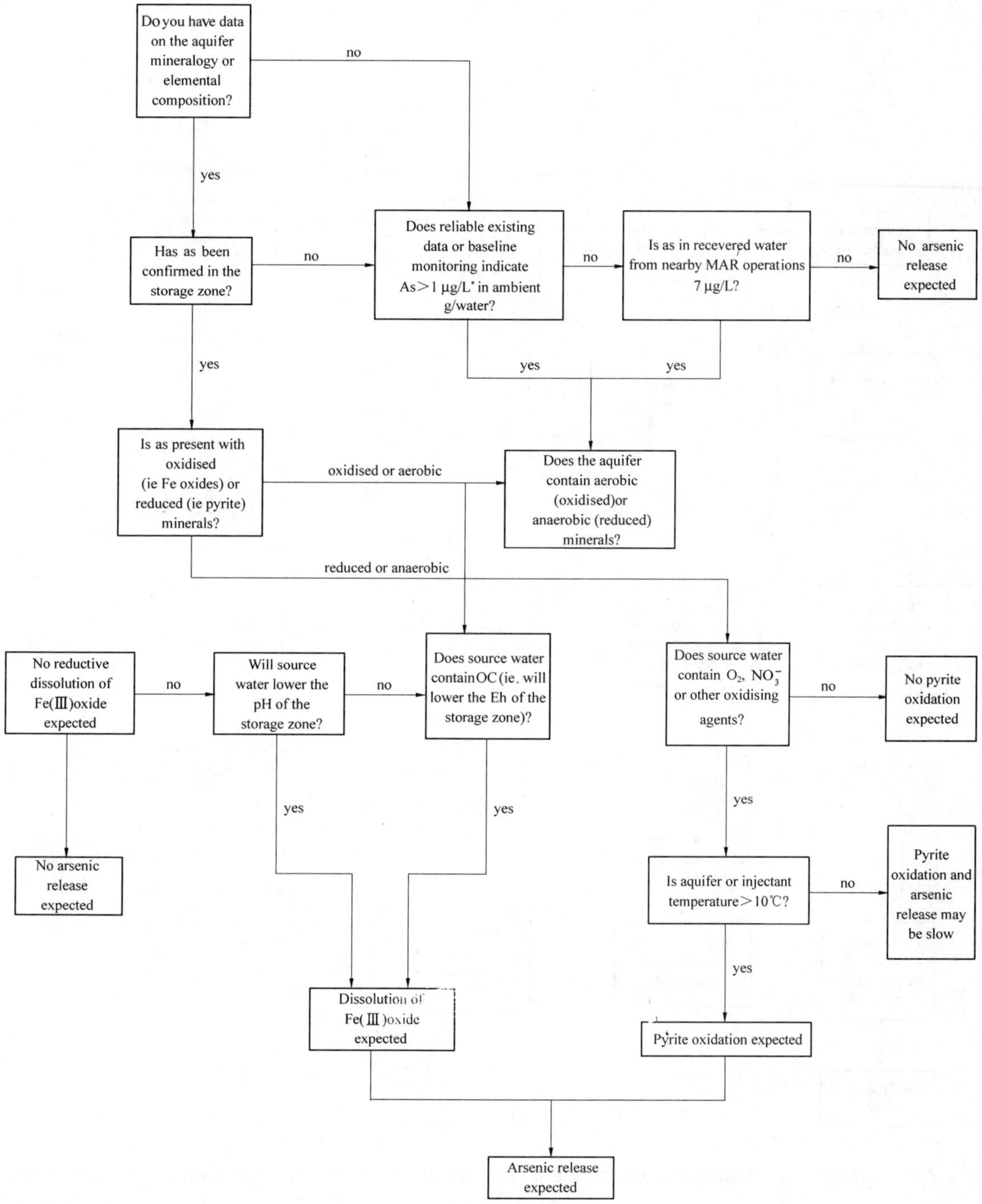

Attached figure 1 Decision tree for identifying arsenic mobilisation in aquifer storage and recharge

*Arsenic (As) concentrations should be reported to 1 μg/L. Other trace metals can be released concurrently.

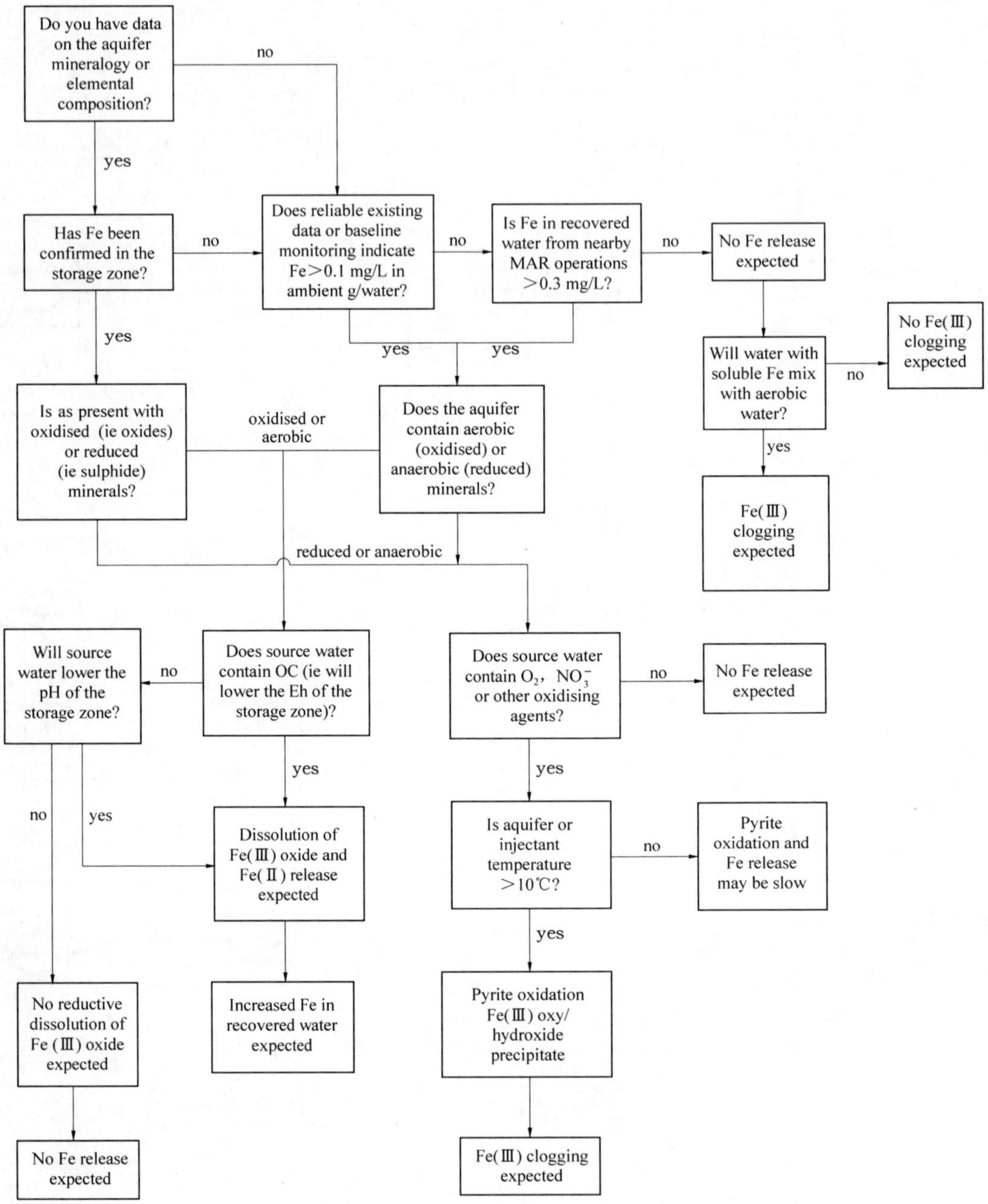

Attached figure 2 Decision tree for identifying potential for iron mobilisation or iron clogging due to redox processes in managed aquifer recharge; iron concentrations may also be affect by mineral equilibrium processes

References

1 Alvarez ME, Aguilar M, Fountain A, Gonzalez N, Rascon O, and D Saenz (2000). Inactivation of MS-2 phage and poliovirus in groundwater. Canadian Journal of Microbiology 46(2):159-165.

2 Amy G and Drewes J (2007). Soil aquifer treatment (SAT) as a natural and sustainable wastewater reclamation/reuse technology: Fate of wastewater effluent organic matter (EfOM) and trace organic compounds. Environmental Monitoring and Assessment 129:19-26.

3 ANZECC-ARMCANZ (Australian and New Zealand Environmental and Conservation Council and Agriculture and Resource Management Council of Australia and New Zealand) (1995). Guidelines for Groundwater Protection in Australia, National Water Quality Management Strategy. ANZECC-ARMCANZ, Canberra.

4 ANZECC-ARMCANZ (Australian and New Zealand Environmental and Conservation Council and Agriculture and Resource Management Council of Australia and New Zealand) (2000a). Australian and New Zealand Guidelines for Fresh and Marine Water. Quality. National Water Quality Management Strategy Paper No 4. ANZECC-ARMCANZ, Canberra.

5 ANZECC-ARMCANZ (Australian and New Zealand Environmental and Conservation Council and Agriculture and Resource Management Council of Australia and New Zealand) (2000b). Australian Guidelines for Water Quality Monitoring and Reporting, National Water Quality Management Strategy Paper No 7. ANZECC-ARMCANZ, Canberra.

6 Appelo CAJ and Postma D (1999). Geochemistry, groundwater and pollution A.A. Balkema Publishers, Leiden, The Netherlands.

7 ARMCANZ-ANZECC (Australian and New Zealand Environmental and Conservation Council and Agriculture and Resource Management Council of Australia and New Zealand) (1994). National Water Quality Management Strategy: Policies and Principles - A Reference Document. Paper No 2, April 1994. ARMCANZ-ANZECC.

8 Arthur J D, Dabous A A, Cowart J B. (2003). Water-rock geochemical considerations for aquifer storage and recovery: Florida case studies southwest Florida, in 2nd International Symposium on Underground Injection Science and Technology, October 22-25 2003, Berkley, California.

9 Bekele, E. Toze, S., Rümmler, Hanna, J., Blair, P., and Turner, N. (2006). Improvements in wastewater quality from soil and aquifer passage using infiltration galleries: case study in Western Australia. Proceedings: 5th International Symposium on Management of Aquifer Recharge (ISMAR5), 10-16 June 2005, Berlin. pp. 663-668.

10 Bekele E, Sklorz S, Douglas G and Prommer H (2007). Batch experiments using reverse osmosis water and sands from the Swan Coastal Plain. CSIRO: Water for a Healthy Country National Research Flagship.

11 Blanc R and Nasser A (1996). Effect of effluent quality and temperature on the persistence of viruses in soil. Water Science and Technology, 33(10-11):237-242.

12 Bouwer H (1978). Groundwater Hydrology. McGraw-Hill Inc. ISBN 0-07-006715-5.

13 Bouwer H (2002). Artificial recharge of groundwater: hydrogeology and engineering.

14 Cavé LC (2000). Geochemical modelling and laboratory batch experiments to test the feasibility of an artificial recharge scheme //Groundwater: Past Achievements and Future *Challenges*, Sililo O *et al.* (eds). Proceedings of the XXX Congress of the International Association of Hydrogeologists, Cape Town, South Africa, 26 November - 1 December 2000. A.A. Balkema, Rotterdam. pp.465-470.

15 Charlesworth, PB, Narayan, KA., Bristow, KL, Lowis, B., Laidlow, G and McGowan, R. (2002) The Burdekin Delta - Australia's oldest artificial recharge scheme //Management of Aquifer Recharge for Sustainability, P.J. Dillon (Ed.) Proceedings of the 4th International Symposium on Artificial Recharge (ISAR4), Adelaide, Sept. 22-26, 2002, Swets & Zeitlinger, Lisse, ISBN. 90 5809 527 4, pp.347-352.

16 Clark JF, Hudson GB and Avisar D (2005). Gas transport below artificial recharge ponds: Insights from dissolved noble gases and a dual gas (SF6 and 3He) tracer experiment. Environmental Science and Technology, 39:3939-3945.

17 Collins KE, Cronin AA, Rueedi J, Pedley S, Joyce E, Humble PJ and Tellam JH (2006). Fate and transport of bacteriophage in UK aquifers as surrogates for pathogenic viruses. Engineering Geology, 85:33-38.

18 Cook PG and Herczeg AL (2000). Environmental tracers in subsurface hydrology, Kluwer Publishers, USA.

19 Dillon P, Pavelic P, Toze S, Ragusa S, Wright M, Peter P, Martin R, Gerges N and Rinck-Pfeiffer S (1999). Storing recycled water in an aquifer: benefits and risks. AWWA J Water, 26(5), 21-29.

20 Dillon, P.J. (2005). Future management of aquifer recharge. Hydrogeology Journal, 13(1): 313-316.

21 Dillon, P. and Toze, S. (eds) (2005). Water Quality Improvements During Aquifer Storage and Recovery. American Water Works Assoc. Research Foundation Report 91056F, 286p + 2CDs.

22 Drewes JE, Sedlak D, Snyder S, and Dickenson E (2008). Indicator and surrogates to assess removal of wastewater-derived contaminants in wastewater treatment and reclamation. Final Report. WaterReuse Foundation, Alexandria, Virginia.

23 EPHC-NHMRC-NRMMC (2008a). Augmentation of Drinking Water Supplies. Phase 2 of Australian Guidelines for Water Recycling.

24 EPHC-NHMRC-NRMMC (2008b). Draft Managed Aquifer Recharge Guidelines. Phase 2 of Australian Guidelines for Water Recycling. Public Consultation Draft, May 2008. http://www.ephc.gov.au/ephc/water_recycling.html.

25 Escalante AEF, Rodríguez MG and Gil FV (2005). Use of environmental indicators in a multi-criteria analysis of the impact of MAR on groundwater dependent wetlands. Proceedings of the 5th International Symposium on Management of Aquifer Recharge (ISMAR5), Berlin, Germany, 11-16 June 2005.

26 Fies MW, Renken RA and Komlos SB (2002). Considerations for regional ASR in restoring the Florida Everglades, USA. In: Management of Aquifer Recharge for Sustainability, PJ Dillon (ed), A.A. Balkema Publishing, Leiden, the Netherlands, pp. 341-346.

27 Fox P (2002). Soil aquifer treatment: an assessment of sustainability. In: Management of Aquifer Recharge for Sustainability, PJ Dillon (ed), A.A. Balkema, pp. 21-26.

28 Fox P, Narayanaswamy K, Genz A and Drewes JE (2001). Water quality transformations during soil aquifer treatment at the Mesa Northwest Water Reclamation Plant, USA. Water, Science and Technology, 43(10): 343-350.

29 Freeze RA and Cherry JA (1979). Groundwater. Prentice-Hall Inc., Englewood Cliffs, New Jersey, p. 604.

30 Gale, I (2005). Strategies for Managed Aquifer Recharge (MAR) in semi-arid areas. UNESCO IHP. 34p. www.iah.org/recharge.

31 Gerges NZ, *et al.*, 2002 South Australian experience in aquifer storage and recovery. In: Management of Aquifer Recharge for Sustainability, PJ Dillon (ed), A.A. Balkema, pp. 453-458.

32 Gibert J and Deharveng L (2002). Subterranean Ecosystems: A Truncated Functional Biodiversity. Bioscience, 52(6):473-481.

33 Glover RE and Balmer GG (1954). River depletion resulting from a well pumping near a river. Transactions American Geophysical Union, 35:468-470.

34 Gordon C and Toze S (2003). Influence of groundwater characteristics on the survival of enteric viruses. Journal of Applied Microbiology, 95:536-544.

35 Greskowiak J, Prommer H, Vanderzalm J, Pavelic P and Dillon PJ (2005). Modelling of carbon cycling and biogeochemical changes during injection and recovery of reclaimed water at Bolivar, South Australia. Water Resources Research 41, W10418, 16.

36 Greskowiak J, Prommer H, Massmann G and Nützmann G (2006). Modeling seasonal redox dynamics and the corresponding fate of the pharmaceutical residue Phenazone during artificial recharge of groundwater. Environmental Science and Technology, 40:6615-6621.

37 Gundersen LCS and Wanty RB (1993). Field studies of radon in rocks, soils, and water, CRC Press, Boca Raton, Florida, 334.

38 Hantush MS (1967). Growth and decay of groundwater mounds in response to uniform percolation. Water Resources Research, 3(1):227-234.

39 Herczeg AL and Dighton JC (1998). Radon-222 concentrations in potable groundwater in Australia. Water 37.

40 Herczeg AL, Rattray KJ, Dillon PJ, Pavelic P and Barry KE (2004). Geochemical processes during five years of Aquifer Storage Recovery. Ground Water, 42(3):438-445.

41 Hijnen WAM and van der Kooij D (1992). The effect of low concentrations of assimilable organic carbon (AOC) in water on biological clogging of sand beds. Water Research, 26(7):963-972.

42 Hodgkin, T. (2004). Aquifer storage capacities of the Adelaide region. SA Dept Water Land and Biodiversity Conservation Report, 2004/47.

43 Howard P (1991). Handbook of Environmental Degradation Rates. Lewis Publishers, USA.

44 Hutchinson A and Randall R (1995). Estimation of injection well clogging with modified fouling index (MFI). Proceedings of the 2nd International Symposium on Artificial Recharge of Groundwater, Johnson

AI and Pyne RDG (eds), Orlando, Florida, July 1994, 710-719.

45 Ibison MA, Sanders FA, Glanzman RK and Dronfield DG (1994). Manganese in recovered water from an ASR well. In: Artificial Recharge of Groundwater Ⅱ, Johnson AI and Pyne RDG (eds), Proceedings of the 2nd International Symposium on Artificial Recharge of Groundwater, Orlando, Florida, 17-22 July, 1994, 539-547.

46 International Association of Hydrogeologists Commission on Managed Aquifer Recharge (2008) home page, www.iah.org/recharge.

47 Jansons J, Edmonds LW, Speight B, Bucens MR (1989). Survival of viruses in groundwater. Water Research, 23:301-306.

48 Jimenez, B. (2003). Health risk in aquifer recharge with recycled water. In: State of the Art Report: Health Risks in Aquifer Recharge Using Reclaimed Water, Aertgeerts R and Angelakis A (eds), Pub World Health Organisation, Report No SDE/WSH/03.08.

49 Jones MA, Harris SJ, Baxter KM and Anderson M (2005). The Streatham groundwater source: an analogue for the development of recharge enhanced groundwater resource management in the London basin. In: Proceedings of the 5th International Symposium on Management of Aquifer Recharge (ISMAR5), Berlin, Germany, 11-16 June 2005, 819-824.

50 Komarova T, Bartkow ME, Müller JF, Carter S and Vanderzalm J (2006). Field evaluation of passive samplers: monitoring Polycyclic Aromatic Hydrocarbons (PAHs) in stormwater. Polycyclic Aromatic Compounds, 26:221-236.

51 Le Gal La Salle C, Vanderzalm J, Hutson J, Dillon P, Pavelic P and Martin R (2005). Isotope evolution and contribution to geochemical investigations in aquifer storage and recovery: a case study using reclaimed water at Bolivar, South Australia. Hydrological Processes, 19:3395-3411.

52 Mackay D, Shiu WY and May KC (1992). Illustrated Handbook of Physical-Chemical Properties and Environmental Fate for Organic Chemicals, Vol Ⅰ-Ⅲ. Lewis Publishers, USA.

53 Martin MW, Xu C and Rattray KJ (2002). Management of physical clogging during ASR at Jandakot, Perth, Western Australia. In: Management of Aquifer Recharge for Sustainability, PJ Dillon (ed), Proceedings of the 4th International Symposium on Artificial Recharge (ISAR4), Adelaide, 22-26 September 2002, Swets and Zeitlinger, Lisse, 199-202.

54 Miller R, Correll R, Dillon P and Kookana R (2002). ASRRI: A predictive model of contaminant attenuation during aquifer storage and recovery. In: Management of Aquifer Recharge for Sustainability, PJ Dillon (ed) Proceedings of the 4th International Symposium on Artificial Recharge (ISAR4), Adelaide, Sept. 22-26, 2002, Swets and Zeitlinger, Lisse, 69-74.

55 NHMRC-NRMMC (National Health and Medical Research Council and Natural Resource Management Ministerial Council) (2004). Australian Drinking Water Guidelines, NHMRC and NRMMC, Canberra. http://www.nhmrc.gov.au/publications/_files/adwg_11_06.pdf

56 NRC (National Research Council) (2008). Prospects for Managed Underground Storage of Recoverable Water, National Academy Press, Washington DC.

57 NRMMC (Natural Resource Management Ministerial Council) (2004). Guidelines for sewerage

systems - Biosolids management. National Water Quality Management Strategy Paper 13. Canberra, Australia.

58 NRMMC-EPHC-AHMC (Natural Resource Management Ministerial Council, Environment Protection and Heritage Council, Australian Health Ministers' Conference) (2006). Australian Guidelines for Water Recycling: Managing Health and Environmental Risks: Phase 1. National Water Quality Management Strategy. NRMMC-EPHC-AHMC, Canberra, Australia.

59 Oliver YM, Gerritse RG, Dillon PJ and Smettem KRJ (1996). Fate and mobility of Stormwater and Wastewater Contaminants in Aquifers, Centre of Groundwater Studies Report No. 67, CGS, South Australia.

60 Olsthoorn TN (1982). The clogging of recharge wells, main subjects, KIWA- Communications 72, 150.

61 Pavelic P, Ragusa SR, Flower RL, Rinck-Pfeiffer SM and Dillon PJ (1998). Diffusion chamber method for in situ measurement of pathogen inactivation in groundwater. Water Research, 32(4):1144-1150.

62 Pavelic P, Dillon P, Martin R, Traegar B and Simmons C (2001). Multi-scale permeability characterisation of a confined carbonate aquifer targeted for aquifer storage and recovery. Proceedings IAH XXXI Congress, Munich, Sept 2001, 'New Approaches to Characterising Groundwater Flow', Vol 2, 859-862.

63 Pavelic P, Dillon PJ and Simmons CT (2002a). Lumped parameter estimation of initial recovery efficiency during aquifer storage and recovery. In: Management of Aquifer Recharge for Sustainability, PJ Dillon (Ed.) Proceedings of the 4th International Symposium on Artificial Recharge (ISAR4), Adelaide Sept. 22-26, 2002, Swets and Zeitlinger, Lisse, ISBN. 90 5809 527 4, pp.285-290.

64 Pavelic P, Dillon P, Toze S, Barber C, Yin Foo D, Knapton A and Jolly PB (2002b). Water banking in the Australian tropics: results from a trial on South Goulburn Island, Northern Territory. In: Management of Aquifer Recharge for Sustainability, PJ Dillon (Ed.) Proceedings of the 4th International Symposium on Artificial Recharge (ISAR4), Adelaide, Sept. 22-26, 2002, Swets and Zeitlinger, Lisse, ISBN. 90 5809 527 4, pp.441-446.

65 Pavelic P, Nicholson, BC, Dillon, PJ and Barry, KE (2005). Fate of Disinfection By-Products in Groundwater during Aquifer Storage and Recovery with Reclaimed Water. Journal of Contaminant Hydrology, 77:351-373.

66 Pavelic P, Dillon PJ, Simmons CT (2006a). Multi-scale characterization of a heterogeneous aquifer using an ASR operation. Ground Water, 44(2):155-164.

67 Pavelic P, Dillon, PJ., Nicholson, BC (2006b). Comparative evaluation of the fate of disinfection byproducts at eight aquifer storage and recovery sites, Environmental Science and Technology, 40:501-508.

68 Pavelic P, Dillon PJ, Barry KE, Vanderzalm JL, Correll RL and Rinck-Pfeiffer SM (2007). Water quality effects on clogging rates during reclaimed water ASR in a carbonate aquifer. Journal of Hydrology, 334:1-16.

69 Pérez-Paricio A and Carrera J (1998). A conceptual and numerical model to characterize the clogging of artificial groundwater recharge systems. Proceedings of the 3rd International Symposium on

Artificial Recharge of Groundwater (TISAR'98), Amsterdam, 21-25 September, 1998, AA Balkema, Rotterdam, ISBN: 90 5809 017 5, 55-60.

70 Pérez-Paricio A and Carrera J (1999). Clogging Handbook. EU Project on Artificial Recharge of Groundwater, Research program on Environment and Climate, Contract ENV-CT95-0071.

71 Popit A, Vaupotic J and Kukar N (2004). Systematic radium survey in spring waters of Slovenia. Journal of Environmental Radioactivity 76, 337-347.

72 Prime Minister's Science, Engineering and Innovation Council (PMSEIC) Working Group (2007). Water for Our Cities: Building Resilience in a Climate of Uncertainty. June 2007. http://www.dest.gov.au/sectors/science_innovation/publications_resources/profiles/water_for_our_cities.htm#publication

73 Pyne, R.D. G. (2005). Aquifer Storage Recovery: A guide to groundwater recharge through wells. 2nd Edition. ASR Press, Gainesville, Florida.

74 Reed DA (2007). Spatial and temporal biogeochemical changes of groundwater associated with managed aquifer recharge in two different geographical areas. PhD Thesis, Department of Microbiology and Immunology, University of Western Australia, Perth, Australia.

75 Reed DA, Toze S and Chang B(2007). Spatial and temporal changes in sulfate-reducing groundwater bacterial community structure in response to managed aquifer recharge. Proceedings of the 6th Conference on wastewater reclamation and reuse for sustainability, 9-11 October, Antwerp, Belgium.

76 Rinck-Pfeiffer SM, Ragusa SR, Sztajnbok P and Vandevelde T (2000). Interrelationships between biological, chemical and physical processes as an analog to clogging in Aquifer Storage and Recovery (ASR) wells. Water Research, 34(7):2110-2118.

77 Runnels DR (1969). Diagenesis, chemical sediments, and the mixing of natural waters. Journal of Sedimentary Petrology 39(3):1188-1201.

78 Sidhu J, Bekele E and Toze S (in press). Concentual Model of the Fate of Pathogens during ASR. In: Water Quality Changes During ASR. AWWARF Project No. 2974, Dillon P and Toze S (eds).

79 Taylor SW and Jaffe PR (1990). Biofilm growth and the related changes in the physical properties of a porous medium I: Experimental investigation. Water Resources, Research 26(9):2153-2159.

80 Toze S and Hanna J (2002). The survival potential of enteric microbial pathogens in a treated effluent ASR project. In: Management of Aquifer Recharge for Sustainability. (ed) P Dillon. Balkema Publishers Australia, 139-142.

81 Toze S, Hanna J, Smith A and Hick W (2002). Halls Head Indirect Treated Wastewater Reuse Scheme. CSIRO Report to Water Corporation, Western Australia, October 2002.

82 Vanderzalm JL, Le Gal La Salle C and Dillon PJ (2006). Fate of organic matter during aquifer storage and recovery (ASR) of reclaimed water in a carbonate aquifer. Applied Geochemistry, 21:1204-1215.

83 Volker, R.E. (ed.) (1981). Proc. Groundwater Recharge Conference, Townsville, July 1980. Aust. Water Resources Council Conf. Series No 3, pp. 60-71. Aust Govt Publishing Service, Canberra.

84 WHO (2006). Guidelines for drinking water quality, 3rd edition.

85 Williams JB, Cowart JB, Arthur JD (2002). Florida Aquifer Storage and Recovery Geochemical Study:

Year One and Year Two progress report. Florida Geological Survey, 131 p.

86 Wolf L, Morris B, Burn S (2006). Urban water resources toolbox: Integrating groundwater into urban water management. Report of EC Project EVK1-CT-2002-00110. IWA Publishing, London.Wall K, Toze S, and O'Hara G (2004). The Fate of Enteric Pathogens in Artificial Recharge Schemes. In Water and Environmental Management. (Ed) Goen Ho, Kuruvilla Mathew. IWA Publication ISBN: 9781843395027, p. 53-60.

87 Wood M, Le Gal La Salle C, Dillon PJ, Hutson J, Fallowfield H and Rinck-Pfeiffer SM (2005). Effect of grain size on biological clogging in porous media leached with synthetic nutrient solutions. In: Proceedings of the 5th International Symposium on Management of Aquifer Recharge (ISMAR5), Berlin, Germany, 11-16 June 2005.

88 Yates MV, Gerba CP, Kelley LM (1985). Virus persistence in groundwater. Applied and Environmental Microbiology, 49(4):778-781.

89 Yahya MT, Galsomies L, Gerba CP, and R.C. Bales (1993). Survival of bacteriophages MS-2 and PRD-1 in Ground Water. Water Science and Technology, 27(3-4): 409-412.

90 Ying GG, Kookana RS and Dillon PJ (2003). Sorption and degradation of selected five endocrine disrupting chemicals in aquifer material. Water Research, 37:3785-3791.

91 Zhou X and Li C (1992). Hydrogeochemistry of deep formation brines in central Sichuan basin, China. Journal of Hydrology, 138:1-15.

1 可管理的含水层补给介绍

1.1 可管理的含水层补给(MAR)的含义

可管理的含水层补给(MAR)是有针对性地对含水层进行回灌，以使回灌水逐渐恢复或取得环境效益。含水层是具有渗透性的地质岩层，能储存水，在自然条件下雨水通过土壤和岩石下渗或溪流渗透补给下面的含水层。人类提高含水层补给的活动可以分为三类：

(1)无意识的含水层补给——例如，在空旷的根深蒂固的植被组成的灌溉区发生的深部入渗，以及城市中雨水管道和污水管道的渗漏。

(2)不可管理的含水层补给——包括雨水排水井和集水坑，及化粪池的沥滤场，这些水通常为没想过要利用的弃之无用的水。

(3)可管理的含水层补给——包括注水井、渗透洼地和沟渠，而回灌的水来自雨水、暴雨水、再生水、自来水和其他含水层的水，这些水最终开采后用于各种用途。

本文将集中阐述第(3)类，还包括为了保护环境和使用开采水，确认由不可管理的含水层补给转化为可管理的含水层补给的时机。

通过可管理的含水层补给提高地下水的补给速率，从而为澳大利亚的城市和农村提供一个重要的潜在水源。本文将介绍可管理的含水层补给的所有形式，但重点是城市地区的应用。

可管理的含水层补给可用于储存各种来源的水，如暴雨水、再生水、自来水、淡化的海水、雨水或来自其他含水层的地下水，这些水在回灌前经过适当的预处理，有时再处理后，可用于饮用水供应，工厂用水，灌溉、冲厕及维持生态系统用水。

实施可管理的含水层补给的原因包括：

- 确保和提高供水；
- 改善地下水水质；
- 防止咸水侵入沿海含水层；
- 减少储存水的蒸发；
- 维护环境流量和依赖地下水的生态系统，从而改善当地的市容、土地价值和生物多样性。

相应的效益包括：

- 通过减少城市排水量改善沿海一带的水质；
- 减少洪水和洪水带来的灾害；
- 促进城市景观的改善，从而增加土地的价值。

可管理的含水层补给将在储存容量中发挥一定的作用，以帮助城市供水同时应付由于气候变化加剧而导致的澳大利亚流域径流的增大，并帮助当前未使用 MAR 的城市地区获得丰富的水资源。

1.2 MAR 的类型

在进行地下水回灌时采用的各种方法要符合当地的各种条件，如图 1.1 和图 1.2 所示。

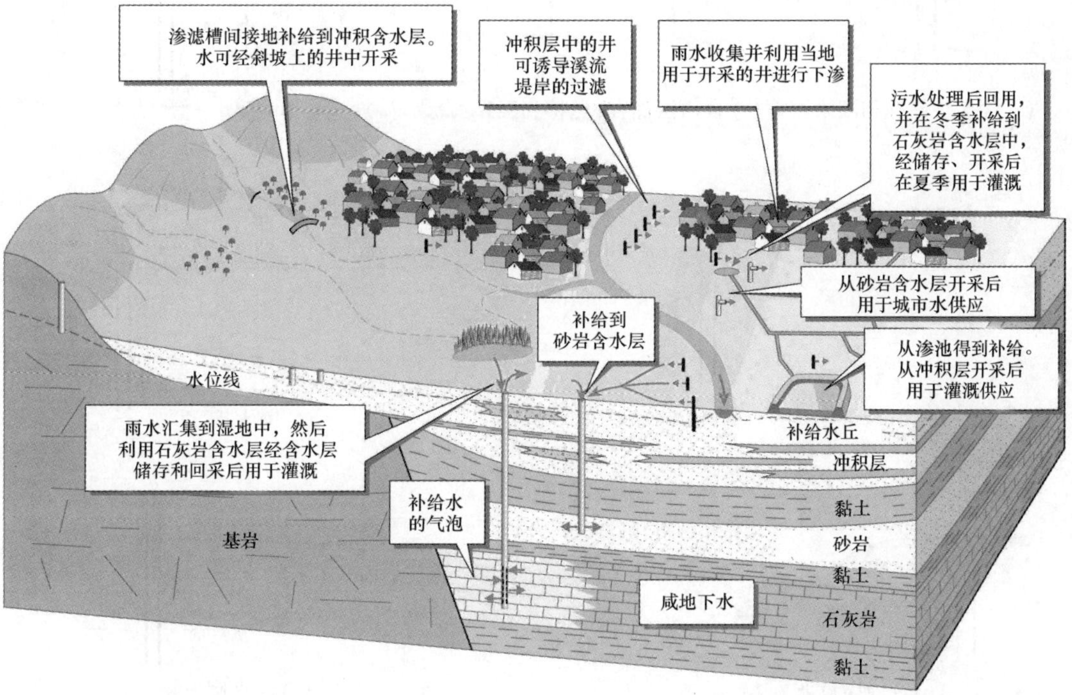

图 1.1 可管理的含水层补给要符合当地的环境，通常受含水层的类型、地形、土地使用状况和开水的使用意图所支配。图中显示了由于补给方法和水源的不同而使用的几种用于储存和开采的不同的含水层，对当地水文地质条件的掌握对于决定采用合适的方法和 MAR 工程的技术可行性是非常重要的。通过井、渗滤槽池和渗透洼地进行回灌(经同意，引自 Gale, 2005)

(1)含水层储存和回采(ASR)：将水回灌到井中用以储存和回采作用的是同一眼井。这在略咸的含水层是非常有用的，这时储存是首要目标而水处理则是一个较小的考虑因素(例如南澳大利亚州帕多克)。

(2)含水层储存、运移和回采(ASTR)：涉及将水注入井中用于储存和回采作用的是不同的井，这种方法通过延长水在两个井之间含水层的停留时间来实现含水层中额外的水处理(例如南澳大利亚州帕勒菲尔德加登斯)。

(3)包气带或干井：在典型的浅井且地下水位埋深大的区域，这种方法允许水质高的水入渗到深层潜水含水层(例如美国菲尼克斯)。

(4)渗滤槽和补给堰：大坝建在季节性的河流上(如只有在降雨或融雪过后才能蓄水的河道)通过滞留水从而渗透到河床，增加潜水含水层的储存量(例如澳大利亚昆士兰州卡利德谷)。

(5)雨水集蓄：将屋面径流运移到井，用沙子或砾石填充的集水井或深井，然后水过滤到达地下水位。通过抽水泵从井中抽取利用(例如西澳大利亚州珀斯)。

(6)堤岸过滤：地下水从位于河流或者湖泊附近的井或深井里抽出来，从而促进地表水体的渗透，再生水的水质也因此得到改善并且更加稳定(例如德国柏林)。

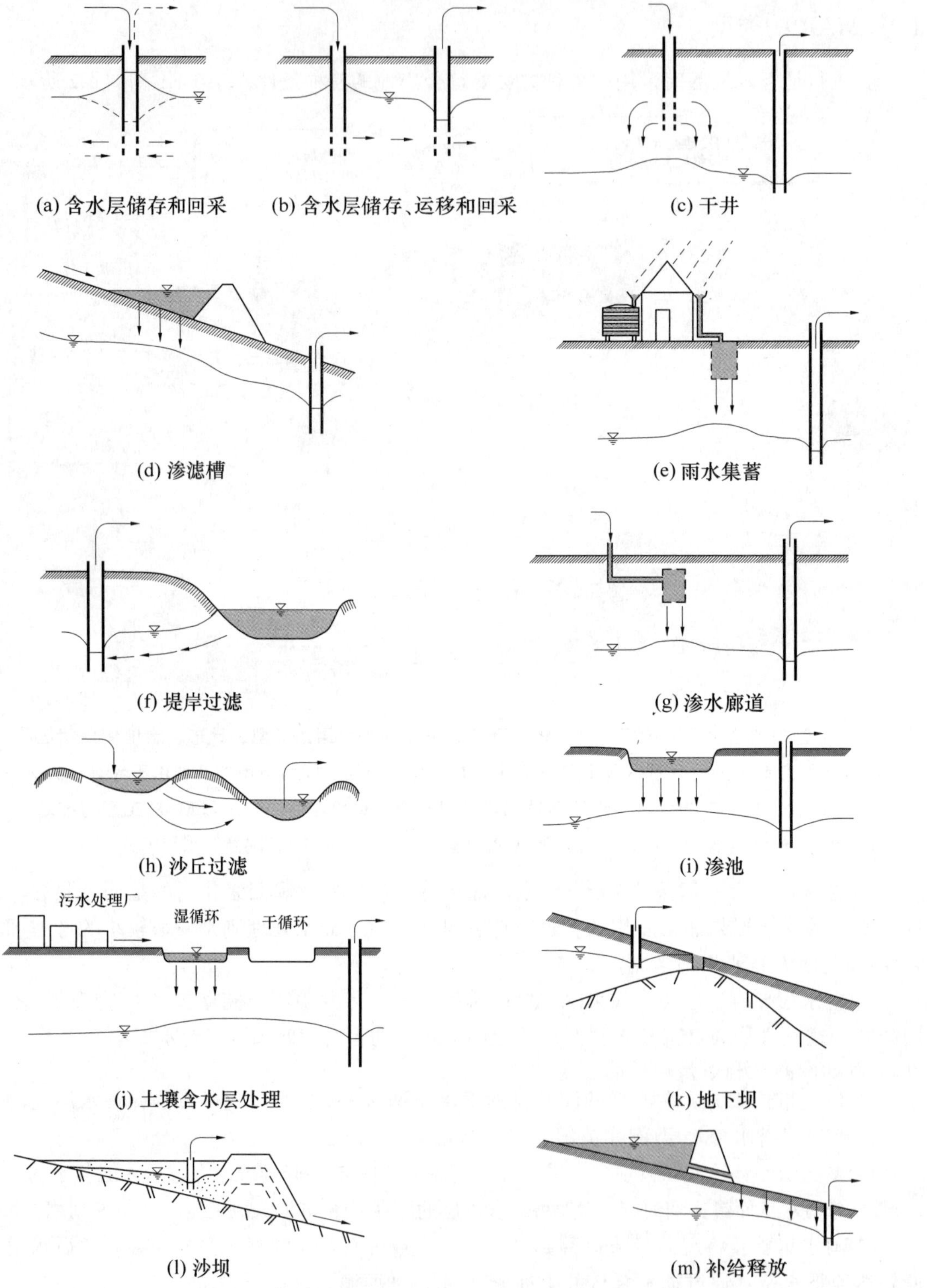

图 1.2　可管理的含水层补给类型图解(资料来源：Dillon,2005, 引自 EPHC-NHMRC-NRMMC, 2008)

(7)渗水廊道：埋入可渗透性土壤的沟渠(包括聚乙烯管或开槽管道)能够允许水渗透通过非饱和带到达潜水含水层(如西澳大利亚州弗罗瑞特花园)。

(8)沙丘过滤：水通过建在沙丘里的池塘进行渗透，然后从海拔较低的井或池塘抽取，通过过滤改善水质，并且能够使供应和需求相平衡(如荷兰阿姆斯特丹)。

(9)渗池：涉及到将地表水运移到河流以外的洼地和渠道，而这些洼地和渠道能够允许水渗入非饱和带到达底部的潜水含水层(例如澳大利亚昆士兰州伯德金部三角洲)。

(10)土壤含水层处理(SAT)：处理过的污水通过渗池间歇性的渗透以促进营养及病原体的移除，这个过程发生在滞留在潜水含水层并通过井而具有开采作用的非饱和带廊道里(例如澳大利亚北部地区艾丽斯斯普林斯)。

(11)地下坝：沟渠建在季节性河流的河床上，水流被高的基准面拦截。底面对于沟渠是非常关键的，其中填充低渗透性的物质。在饱和冲积层能够帮助储存洪水为储存和住宅所用(这种例子在巴西东北部很普遍)。

(12)沙坝：沙坝建立在干旱地区且处于低渗透性岩性的季节性河流的河床上。当水流或持续的洪水发生时沙坝就会阻挡形成沉积物，然后被抬升形成一个“含水层”，在枯水季节可以开采井加以利用(例如肯尼亚基图伊)。

(13)补给释放：建立在季节性溪流上的大坝可以用来滞留洪水，用途包括缓慢的释放水进入下游的河床并渗透到下面的含水层，从而大大提高补给量(例如南澳大利亚小第河)。

可管理的含水层补给的地点和方法的选择取决于水文地质、地形、水文条件和区域的土地利用情况。由于共同的物理属性，在同一地区发现相似的可管理的含水层补给工程类型是常见的。而对不同的地区，方法则可能大不相同。

为提高水的供应，可管理的含水层补给技术在许多国家得到广泛应用，在半干旱和干旱地区尤为显著，而在湿润地区，该技术主要用于改善水质。国际水文地质学家协会有一个关于可管理含水层补给的委员会其网站包含了许多案例研究，见其文件下载(www.iah.org/recharge/).

1.3 MAR 工程的组成

图 1.3 显示了所有可管理的含水层补给工程类型通常所包含的 7 个要素。但是由于可用于储存的含水层类型不同，可管理的含水层补给工程可能会大不相同。主要有两种含水层类型：一种是在低渗透性岩层之间，这时可管理的含水层补给需要向井中注水来进行回灌(见图 1.3(a))；还有一种是非承压的且允许水通过可渗透性的土壤下渗，这时可以通过洼地和沟渠进行补给(见图 1.3(b))。

参数和大小的选择依赖于：

- 低渗透性岩层的厚度；
- 所必需的下渗速率；
- 可利用的土地和花费；
- 和其他土地利用协调；
- 交通便利；
- 避免虫害甚至鸟类的袭击(例如在飞机场)。

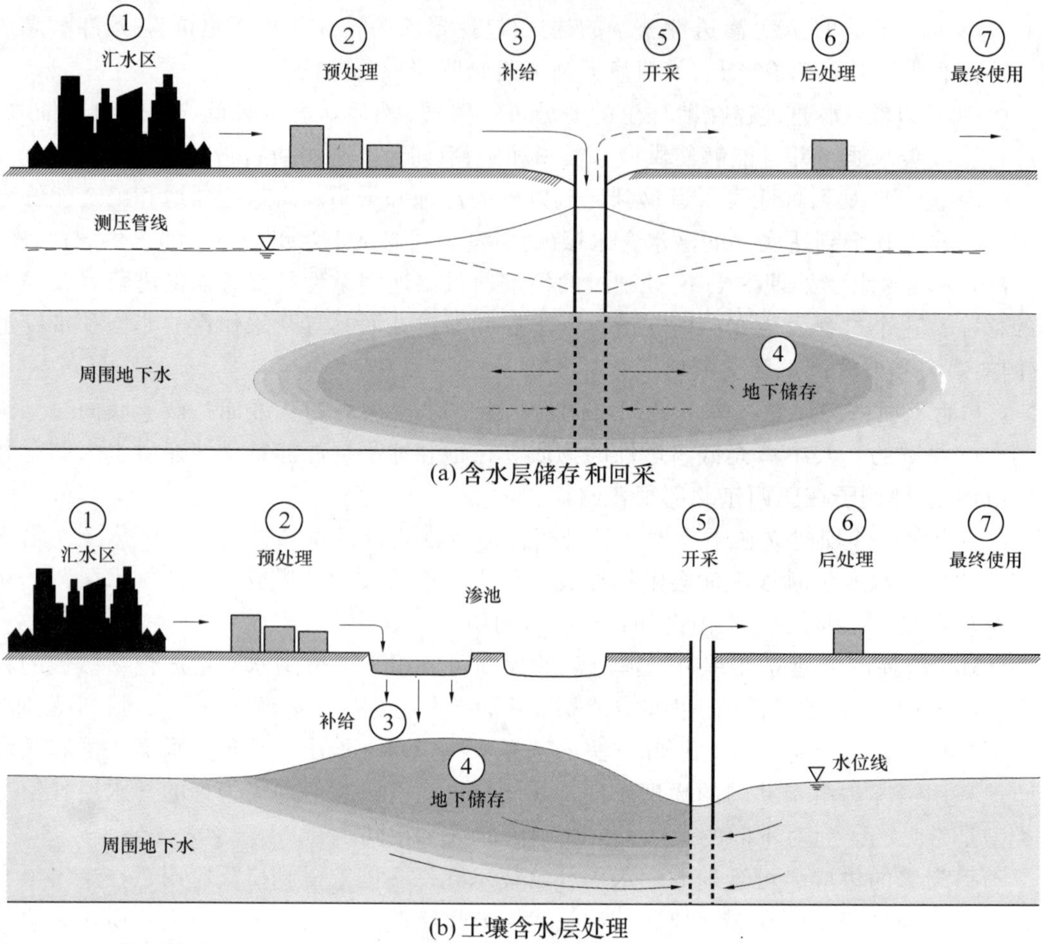

图 1.3　可管理的含水层补给的两个例子(图 1.3 中显示了在每个系统中通常包括的 7 个要素。测压管水位就是井建好后井中的水位线。对于潜水含水层这是水位线。在补给时水位线上升；在回采井周围水位线下降，图 1.3 中的数字及相应的含义见图 1.4)

一些回灌方法依赖于特殊的地貌特征，例如：

- 沙丘和洼地 (主要在暴风雨或融雪水时，低压能够带动水)；
- 位于淤积河流之下浅的岩床或堤坝；
- 在冲积而成的山谷中限制河流对过滤池或沙坝的渗透。

在城市地区，土地价格通常很高，因此通常采用土地利用率高的方法(如含水层储存和回采或土壤含水层处理)，这种方法依赖于含水层是非承压的还是承压的；或是否有可用于储存功能的含水层。在农村地区，土地价格较低，因此回灌的水量很大时采用渗池和土壤含水层处理方法是最划算的。

在选择方法时原水的水质也起到重要的作用。一般而言，若原水浑浊、营养物质浓度高或者不稳定，用井进行回灌很可能导致立即堵塞。在这种情况下，渗透洼地需要定期清洗。作为选择，如果需要较高的处理水平的话，对反清洗井中的沉积物和微生物的清除，以及水处理后副产品的排放或循环都要进行防备。

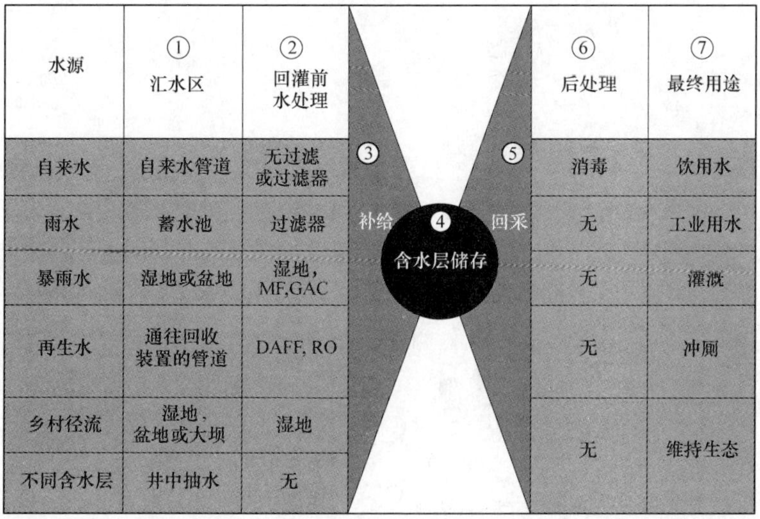

图 1.4　可管理的含水层补给的水源类型，汇水区的方法和预处理在回灌前所有水源通过合适的处理措施，经过开采后可用于任何一种用途(圈中元素对应的数字及含义见图 1.3)。给出的所有补给方法见图 1.2。MF 表示微生物过滤，GAC 表示颗粒活性炭过滤，DAFF 表示溶气气浮过滤，RO 表示反渗透

在选择回灌的方法之前所有这些因素都要进行考虑。对于研究的花费，工程所涉及的潜水含水层系统比承压含水层系统要多。因为除含水层的特性外，还要掌握穿过水位线的土壤的剖面。在一些区域，当使用不同的含水层或者不同的回灌方法时，储存水的最终用途是不同的。

水源、水处理和最终用途有很多结合形式(见图 1.4)。通常水质较差的水源在回灌前需要进行较高水平的处理措施，除以下几方面：

- 含水层已经含有水质高的水；
- 开采后的水有较高的使用价值，如作为饮用水；
- 含水层是由细颗粒组成的，因此对于补给盆地、沟渠或井应该避免经常性的或永久性的阻塞。

对于城市暴雨补给略咸的石灰岩含水层，采用被动的处理方式是合适的，例如湿地可能被开采用来灌溉，这不需要后处理。如果开采后用来作为饮用水则需要水在含水层有一定的停留时间以确保水质，在细颗粒组成含水层的地点人们发现需要微生物过滤(MF)和颗粒活性炭过滤(GAC)以防止 ASR 井堵塞。这些要求比保护地下水水质和开采后水的合理使用更为迫切。如果水浑浊度低可能没有必要进行处理，河水可以通过渗透洼地来提高灌溉供应。如果再生水回灌后作为饮用水供应，则在回灌前需进行深层处理，这即是保护健康和环境，以风险为本的 MAR 新方针的简要概述。

1.4　澳大利亚在 MAR 方面的经验

在最近几年里，澳大利亚可管理的含水层补给的多样性和地理分布数量逐渐增加。2008 年，澳大利亚的 5 个州的区域内已有运行的可管理的含水层补给工程，2 个州正在进行研究(见图 1.5)。

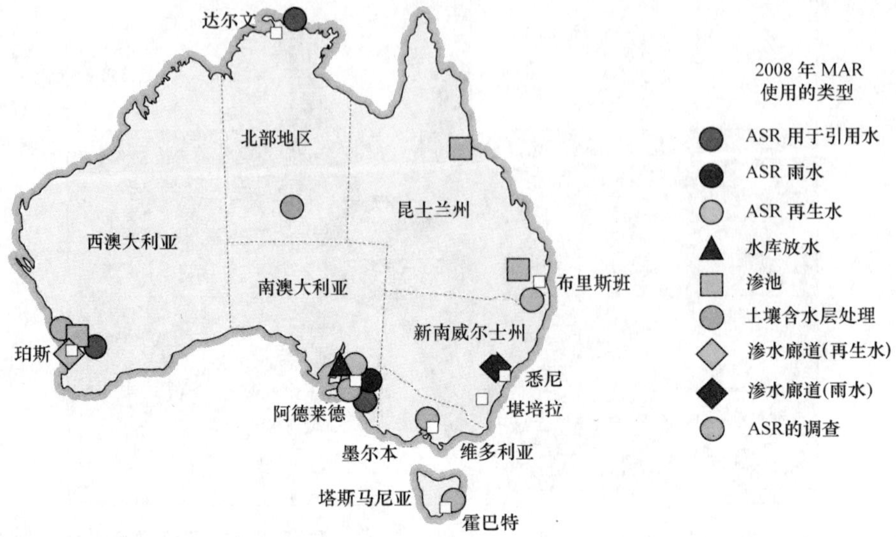

图 1.5　2008 年澳大利亚地区可管理的含水层补给的位置及类型

在澳大利亚移民至城市中心以前，可管理的含水层补给首先开始于农村地区。预计当水资源管理帮助恢复环境流量后，可管理的含水层补给会再次在农村地区发展。

渗透洼地于 20 世纪 60 年代中期建立在昆士兰州的伯德金郡三角洲这是澳大利亚服务时间最长和当前最大(达到 45 GL/a)的可管理的含水层补给设施。(Volker, 1981; Charlesworth 等, 2002)。

在 20 世纪 70 年代，回灌堰建立在卡利德和 Lockyer Creeks，通过井进行补给出现于 1970 年在南澳大利亚州 Angas-Bremer 的灌溉区，到 1992 年回灌井的数量增加到 30 眼且补给量达到 2.4 GL/a (Gerges 等, 2002)。补给释放(1.5 GL/a)开始于 1979 年南澳大利亚小第河 (Dillon, 1984)，目的是代替自然条件下河流对北阿德莱德平原含水层的补给。

在 1992 年，城市暴雨水的含水层储存和回采最初在位于南澳大利亚州石灰岩地区的 Andrews 农场开始发展；后来 1994 年在处于断裂岩的 Regent 花园建立(Gerges 等，2002)；1999 年，再生水的含水层储存和回采开始于南澳大利亚州的玻利瓦尔；2000 年西澳大利亚州 Halls Head 开始通过渗池进行储存和开采(Toze 等, 2002)；而在 2005 年西澳大利亚州简达科特则通过渗渠(简达科特等, 2006)进行储存和开采。在 2000 年，利用自来水进行含水层的储存和回采在西澳大利亚简达科特开始应用 (Martin 等, 2002)，而在 2001 年在澳大利亚北部地区 Warruwi 利用较浅含水层的水进行储存和回采(Pavelic 等, 2002)。在 2007 年，利用暴雨水回灌的渗渠建立于新南威尔士的肯辛顿；在 2008 年，再生水的土壤含水层处理开始在澳大利亚北部地区的艾利斯斯普林斯发展，同年关于暴雨水的含水层储存和回采在墨尔本、维多利亚的几个地点正在进行研究。

位于南澳大利亚州甘比尔山的暴雨处理井自 19 世纪 80 年代开始运行，在 19 世纪 90 年代被验证有助于城市的供水(原先来自蓝湖)。后来 Wolf 等在 2006 年创建的风险管理设计被采纳，将不可管理的含水层补给转变为可管理的含水层补给。在珀斯，对于家庭灌溉和小部分的自来水供应井，屋面径流和暴雨流入深井和洼地的补给对利用井进行开采做了实质性的贡献。风险估计的提出将以前的不可管理的补给转变成可管理的含水层补给。

1.5 MAR 的目的

在农村地区，可管理含水层补给主要是增加地下水灌溉供应的安全性和提高灌溉水的水质。在城市地区进行可管理的含水层补给有很多目的(见图 1.6)。例如，在索尔兹伯里，只是用以减轻洪水，当改善城市市容和沿海水质的湿地建成后，含水层储存和回采才能被考虑。注入和开采滞留水的额外费用给供水和自来水的价格带来了竞争力。

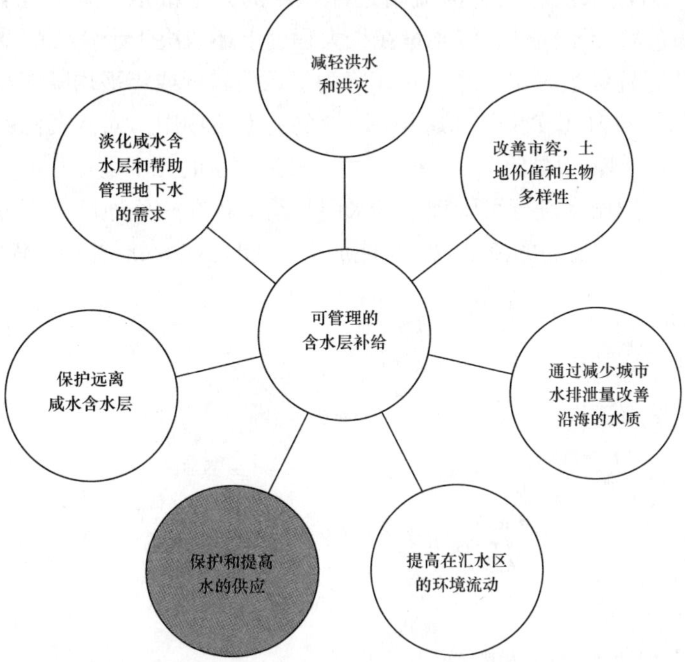

图 1.6 可管理的含水层补给工程，尤其是在城市地区，除供水之外还有许多目标，而这些目标随地点的不同而变化。所有关于经济、社会和环境效益以及项目成本的因素都需要在选择项目时加以考虑

对于新项目，在任意地点引入可管理的含水层补给可能有多种原因。综合利益如下所述，不仅仅涉及供水，也可能决定引入可管理的含水层补给的项目能否收益。

在确保和加强饮用水供应的特殊情况下，在澳大利亚引导建立可管理的含水层补给工程以快速满足当地政府、体育俱乐部和产业部门的非饮用水在商业上的需要。

当考虑多目标的效益和成本时，与其他替代技术形成对比的是各国政府和水利事业对可管理的含水层补给成本和可靠性的信心越来越大。虽然紧急供应具有最高的社会价值，但如果不建立程序来支持这方面的投资及保护战略储备，城市地区含水层的存储能力会被用于商业用途。这需要采用一种分层方法来分配含水层的存储空间和收集的雨水，首先确保战略储备和替代水供应。如有剩余容量的话，可分配给新的供应。这将有助于使城市供水安全的水资源价值发挥最大效益，尤其是考虑到气候变化和人口增长。

1.6 气候多变性成为 MAR 的一种动力

将来，由于气候变化，许多大洲在季节动态中将会面临年平均降雨量的减少、更加频繁的高强度降雨、高温和较高的蒸发速率等问题。因此，为保持当前城市的用水安全，将需要更加稳定的替代品、更多的水储藏量和一系列需求管理措施。

气候变化也将影响土地利用、土壤覆盖、水土流失和火险频度，所以流域内传统供水的水质可能会变得更加易变，可能需要进行水处理。因此，城市地区传统集水区水质的优势很可能随着时间的推移而部分退化。

作为一项适应气候变化的措施，目前在城镇堤岸过滤可以通过直接利用河水而被采用。可管理的含水层补给的应用在欧洲是常见的，作为预处理及消除流出水水质不稳定的一项措施，这里饮用水供应是从河流附近冲积层的井里抽取，而不是从河流里抽取的。

储存的需要量除跨年不同外，降雨量在很大程度上还取决于水资源的季节变化。长期干旱的城市(图 1.7 中，连续最干旱 6 个月比年降雨量数值较低或年平均降雨量比蒸发量数值较低的城市)，与降雨量均匀且较湿润的城市(图 1.7 的右上角)相比，对水资源的储存有更大的需求量。然而对于非常干燥的城市(图 1.7 最左边)，收集雨水的几率是很小的，又由于投资资金的低使用率，所以利用雨水进行可管理的含水层补给工程在这些地区是不可行的，以这些地方的再生水作为一个非常稳定的流量，很可能成为可管理的含水层补给工程的首选水源。

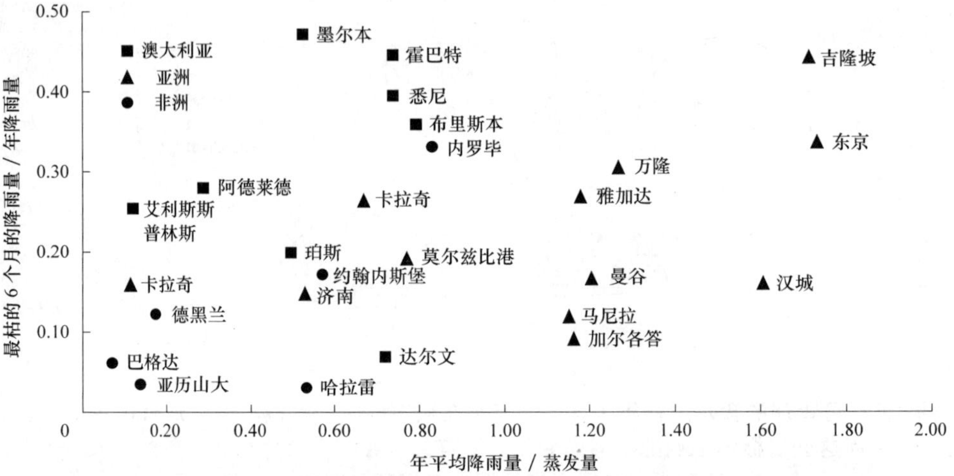

图 1.7　降雨量的季节性以连续最干旱的 6 个月降雨量与年降雨量的比以及降雨量和蒸发量之比的形式表达出来，显示了季节间水储存的需求和利用雨水满足水需求的可行性

在较湿润的城市有更多可利用的雨水，这样减少了收集雨水的单位成本。因此，从气候角度来看，澳大利亚城市达尔文、布里斯本、悉尼和霍巴特在暴雨水收集方面更有前途，但在提倡可管理的含水层补给方面却意外的比阿德莱德、珀斯和墨尔本积极性低一些。这种情况的原因，可能是涉及到这些城市地下的假定的含水层属性，这个主题将在后面讨论。

1.7　可选择水资源供应相应的 MAR 成本

作为水资源循环的一部分，可管理的含水层补给利用暴雨水或再生水回灌后，通常用于直接饮用水或非饮用水，以替代现存的全部或部分总管道水资源供应。在任何城市、城镇或乡村地区，可管理的含水层补给的成本和收益都可以与改良水的保护、发掘新的地表水供应或含水层、雨水储水池以及地下水或者海水淡化等项目的花费及收益来进行比较。当地的地域状况决定了各项目的成本，而各地区间任意项目成本和不同项目间的相对成本效率也存在较大差异。

图 1.8 显示了系列单位成本，这些单位成本的评估是针对 4 个城市在连续的基础上进行的，该评估可用于一系列投资项目。可管理的含水层补给可能在储存、附加被动处理以及三种方法(暴雨水再利用、间接饮用水再利用和非饮用水再循环)的分配中发挥作用。

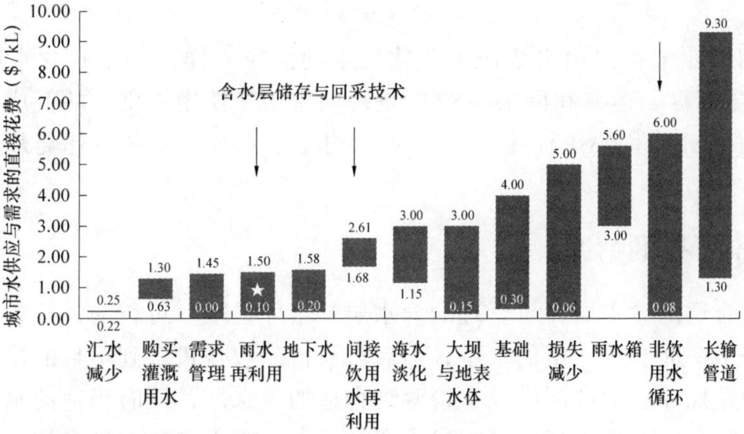

(来源：Marsden Tacob 分析基于悉尼、珀斯、纽卡斯尔供水计划进行的，间接饮用水再利用的低约束力评估是基于城市图文巴进行的)

图 1.8 澳大利亚 4 个城市供水改进和需求减少方面直接费用的比较

(来源：PMSEIC ，PMSEIC, 2007, 城市用水状况，各条带代表各项成本的变动范围)

考虑每项新资源或保护水措施的规模、质量、可靠性、接近度和成本，大部分城市将寻求多样化的水源以加强水资源供应的安全性。在发掘新的水源之前，需要评估各资源的属性，并在采取确保未来的水供应的综合措施之前，考虑到气候变化、人口增长、改变人口统计学、现有基础设施及其寿命，以及环境和社会措施，制订出一个综合的投资计划。

图 1.9 显示了在考虑到已建立的资源和需求增长导致的供水安全性降低的条件下，该一

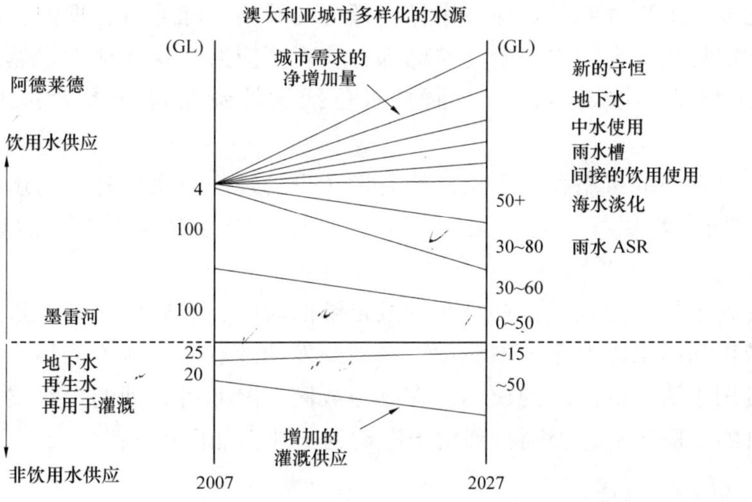

图 1.9 多元化的水源可能包括将可管理的含水层补给作为未来城市供应品之一，以帮助其他项目，如海水淡化，扩大储存量和供应量

系列项目对一个城市可行的案例。三角形区域代表有潜在发展空间用以完善供水能力投资的项目。最终各项目的利用程度取决于每种供选择资源的规模以及开发时的财政费用。例如，海水淡化不受容量约束且亦不由降雨量决定，并且建造速度相对较快，但却是水资源供应开发中花费最多的项目。因此，除满足现实需求外的扩充依赖于供应和需求间的缺口。

与灌溉和工业中将饮用水替换为非饮用水的供应一样，暴雨水含水层储存和回采(ASR)与含水层储存、运移和回采(ASTR)是一项应用于饮用水供应的低成本项目，再生水的含水层储存和回采(ASR)将被应用于增加再生水水量，园艺、葡萄栽培以及城市非饮用水用途。

1.8　MAR 不可行的情况

只有存在合适的含水层，可管理的含水层补给才是唯一可行的，换句话说，人们可以利用足够补给率而获得充足的水量，然后将获得的收益投资到项目的建设中去。如果在负担得起的钻探深度的范围内没有检测到合适的含水层，同时当地的水文地质研究显示在水源附近无合适的含水层或没有对水资源的充分需求来改善其经济，那么可管理的含水层补给是不可行的。

如果井释放的水量小，发育能力可通过计算补给或开采的最低经济速率来确定。补给井或开采井的数目，或进行补给所要求的补给速率或开采的渗水廊道或渗池的面积要与工程的预算进行比较。一般来说，如果含水层中井的排泄量很小，那么为避免堵塞通常处理补给水的成本就很高；甚至开采单位体积水的维护费用会更高。

如果环境风险不能通过经济上可行的预防措施减少到可以接受的水平，则不推荐采用可管理的含水层补给(考虑到工程所有的成本和效益)。如果奖励机制和管理效率低，且对倡导者只有少量利益的话，则不鼓励进行勉强可行的项目。

处于潜水位的非承压含水层也可能会出现问题。没有对建筑物和其他基础设施造成水涝、盐碱化或岩土工程问题的地方来储存额外的补给。除非可管理的含水层补给工程将增加水的抽取量作为提议中一个完整的部分，否则应避免这些地方。然而，这些位置的管理在很大程度上被限制住了，而且水位线的波动很可能大大超出自然条件的波动。

如果一个含水层的可用存储容量已经有助于可管理的含水层补给的运行，那么进行额外的可管理的含水层补给是不切实际的，并且可能会对现有工程的运行产生负面影响。

可管理的含水层补给也需要补给水源及足够的时间，以确保可开采量来补充项目的设立费用。城市暴雨水需要土地来滞蓄。在新开发的地点，土地价值的增加是由于水资源的收益超过用于滞蓄的土地的成本。在已建成区，除非可管理的含水层补给是和谐的市区重建项目的组成部分，否则找到用于滞蓄的土地更加困难。

2 怎样用风险评估框架构建 MAR 项目

潜在的倡议者首先需要知道在进行 MAR 项目之前是否具有以下所包含的五个基本要素(第 1 阶段，评估)。如果该项目在第 1 阶段是潜在可行的，则 MAR 指南草案(EPHC-NHMRC-NRMMC, 2008b)将引导倡议者从调查(第 2 阶段)和运行试验(第 3 阶段)到实施中的项目(第 4 阶段)。

2.1 阶段 1——五个基本要素

成功的 MAR 项目包含五个重要要素(见图 2.1)：

- 对开采水有足够的需求；
- 有充足的补给水源；
- 有合适的含水层来储存和开采水；
- 有足够的土地来收集和处理水；
- 有效的管理工程能力。

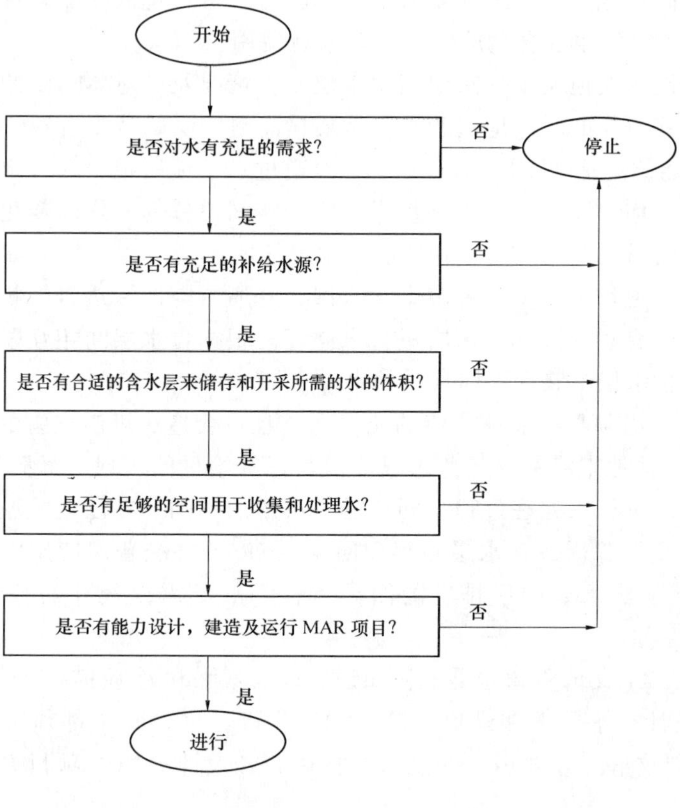

图 2.1 考虑是否承担可管理含水层补给项目的程序

(1)要求：对于 MAR，补给所产生的开采水需求量(一定经济尺度范围内)或明确的环境效益要求必不可少。开采水的用途也需要加以界定(见图 1.4)。通常这会提供税收收入来支付工程中水的供应成本。在城市地区需要滞留雨水来减少洪水、改善沿海或受纳水体的水质、改善城市市容和提高土地价值，也会有助于税收收入，对于再生水工程，处理过的污水流到海里的排放量的减少可为 MAR 的投资提供动力。

(2)来源：对于补给需要确保使用水的权利。年平均补给量应超过年平均需求量，用剩余的部分来建立一个缓冲带使储存达到稳定和水质的要求。在超额汇水区很难确保地表水使用权是可用的。

通常城区的雨水和再生水资源丰富，但在利用前需进行处理和储存。在澳大利亚城市多年平均雨水排放量是自来水使用量的 85%~145%，污水排放量是自来水使用量的 50% ~ 80%。

为城市供水作出有益贡献的雨水或再生水的可用性不是一个制约因素。对雨水收集和使用来说，主要限制是储存来自径流的水的能力，这些水随后作为饮用水供应或灌溉，工业用品或其他非饮用水用途。在城市地区 MAR 可以提供一种储存水的经济手段。

在安置任何坝或含水层之前，污水排放需要进行广泛的处理。含水层在进行水的被动处理时有优势，并且在开采饮用水之前允许更长的滞留时间。

(3)含水层：对于 MAR 来说，一个合适的含水层是至关重要的，它需要有充足的补给速率、足够的储存容量和用来恢复滞留水的能力。可首选低盐度并且接近咸水的含水层，因此其与补给的淡水混合后应仍容许开采后使用。

需要了解当地水文地质条件来识别含水层发育情况及其对 MAR 的适用性。新井在施工之前，现有井所在的基础地层和水文地质信息是宝贵的背景资料，这些资料可以用来确定 MAR 可能的地址(见图 2.2)。一般可以通过水文地质报告在一定程度上了解当地含水层的情况及其均一程度。由于含水层性质随空间位置的变化而变化，因此一般来说由一个地点来推断附近地点的发育和性能是不可靠的。

任何特定位置上均可能发育有多个含水层，并通常与低渗透岩层相间分布。其容许选择一个或多个最有利特性的含水层进行水储存。根据含水层间相互联系的程度，可能在同一地点不同含水层中储存不同水质的水。

(4)滞留存储：必须具有能够滞留的充足水体且不会造成洪涝灾害的开阔空地，或大坝、湿地、池塘、盆地来滞留充足的水以确保目标补给量的实现，而不会造成洪涝灾害。即使后来确定要求进行水处理过程，同样需要可用的空间。在已建的城市汇水的空间是雨水收集的一个主要障碍，含水层储存和回采(ASR)的井被普遍使用以避免渗透系统的土地要求。对于从污水处理厂排出的循环水，一般不需要额外的补给设施来滞留储存。

(5)管理能力：水文地质和岩土工程知识，以及蓄水和处理工艺设计、水质管理、对水敏感的城市设计、水文学和模拟、监测和报告等方面的知识都须符合管理要求。在第 2 阶段将需要这些专业知识，并且需要在调查和设计 MAR 项目方面经验丰富的顾问。

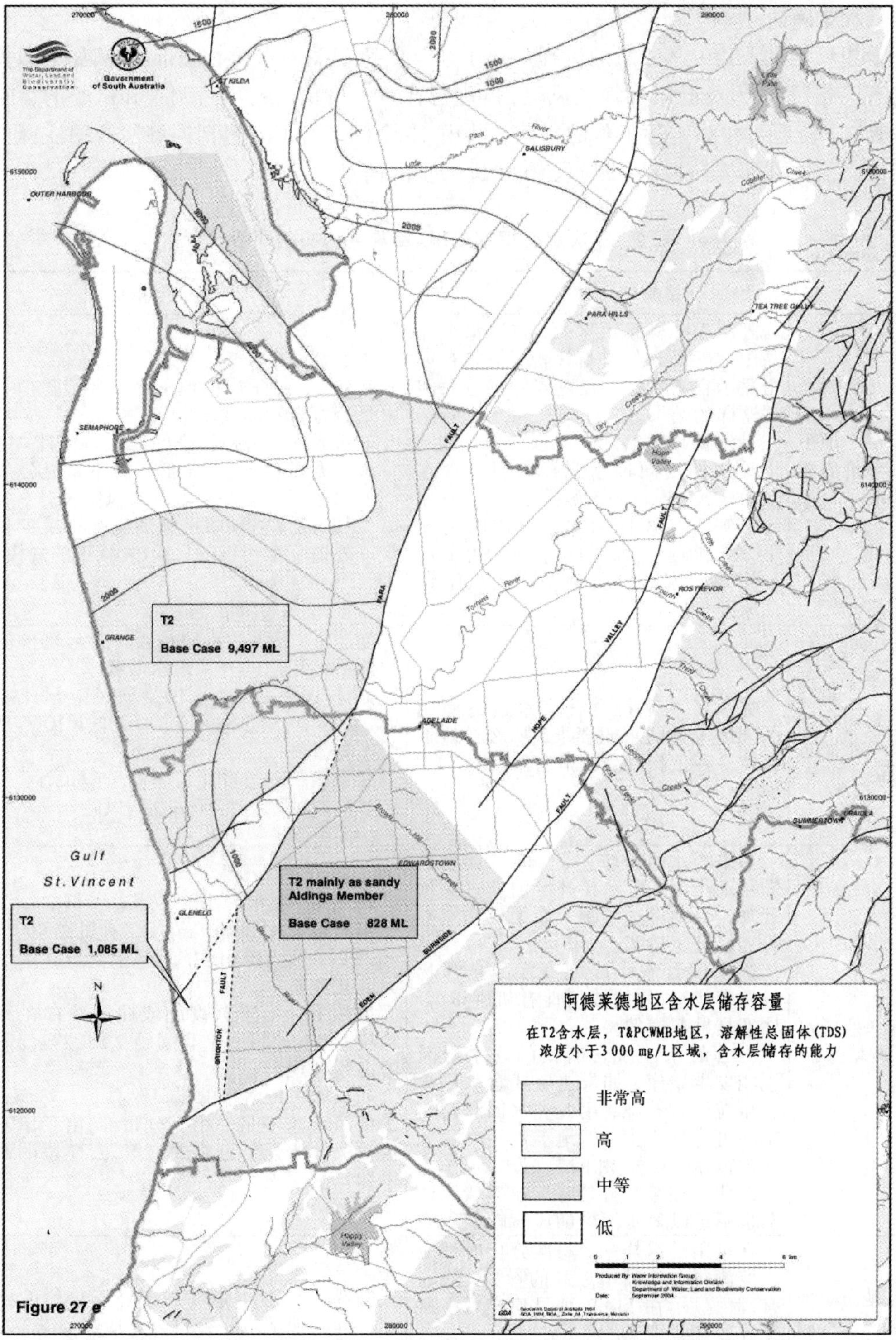

图 2.2 阿德莱德中央部分的 T2 含水层 ASR 发展潜力(from Hodgkin, 2004)(图中不同阴影代表基于区域的水文地质条件下的 ASR(含水层储存与回采)项目的发展前景，主导因素是潜在的注入率，其可以通过记录井和历史的水文地质报告得到)

2.1.1 难度确认

设想项目的难度从表 2.1 的流程进行了解。依照相关的水质管理准则(例如，国家水质管理策略)管理人类健康和环境风险，需进行耗费大量精力的项目调查和试验的运转工作，表 2.1 流程可作为上述工作的指南，共由 13 个问题组成，包括回答所需的注释信息和结果，需要在研究阶段(第 2 阶段)获取进一步的信息。

表 2.1 决定设想 MAR 项目难度的实例评估

评估所需的信息		问题及难度指标
1. 关于地下水环境价值源水的水质	凡多重抽样是可用的，应取每个样品中浓度最高的进行评价，除非有理由显示在 MAR 工程建成后将避免出现这样的结果 另外，缺乏实际水源水质数据时，可采用使用同种水源补给同一含水层的 MAR 项目中的数据 在缺乏上述任何数据资源的情况下，可利用一般的数据。质量可变性和影响因素评价要遵循最大风险评估原则	问题 1：对于周围地下水的环境价值，源水是否符合水质要求？ 如果是——进行污染的低风险预计。这是一个必要条件，但对于低风险不是充分条件 如果否——很可能需要高风险预计，期望第 2 阶段评估预防措施的调查将减少衰减带以外地下水污染的风险(衰减带的规模)
2. 关于最终用于环境价值的再生水源水的水质	如果源水不符合开采水最终使用的环境价值的水质要求，那么在地下存在一种依赖衰减的危害	问题 2：源水的水质是否符合环境价值的预期最终用途的开采水水质要求？ 如果是——开采水污染低风险预计。然而，由于含水层的反应，对于低风险不是一个充分条件 如果否——很可能需要高风险预计。预计第 2 阶段的调查将评估这一风险
3. 关于源水水质与堵塞	如果源水水质较差，土壤或含水层颗粒细小时，源水在补给前要进行预处理，否则渗透洼地、渗渠或回灌井可能会发生堵塞 当水中含有中度或高度的悬浮物或营养物质(如氮或活性有机碳)时，堵塞是最普遍的 当含氧水注入含有铁的含水层时也能发生堵塞。如果土壤或含水层含有粗粒或大孔隙，用此类水回灌时堵塞的可能性不大，但地下水污染的风险却很高(问题 1 和问题 2 已覆盖) 即使是在细粒介质中，缺乏堵塞证据也不足以显示污染的风险较低	问题 3：源水的水质是否差，例如： 当总悬浮固体>10 mg/L，有机碳总量>10 mg/L，总氮>10 mg/L 时，土壤或含水层有无大孔隙？ 如果有——渗透设施或回灌井存在高风险堵塞。不管问题 1 和问题 2 的答案都需要考虑预处理 如果否——堵塞的低风险预计。然而，对于低风险这不是一个充分条件，由于堵塞所依赖的含水层特性将通过第 2 阶段的调查得知
4. 地下水水质与再生水最终用于环境价值的关系	在可用的采样中，每次分析时都应当监测每个样品中最高的参数；除非有理由显示在 MAR 项目建成后将避免导致这些结果产生 另外，在被提议的地点若缺乏地下水水质的数据，则可使用同一含水层中周围井的数据	问题 4：对于最终用于环境价值的开采水，周围地下水是否符合水质的要求？ 如果是——不足开采率的低风险预计 如果否——不足开采率的风险预计

续表 2.1

评估所需的信息		问题及难度指标
5.地下水与饮用水水质	含水层的环境价值需要得到有关当局的详细说明。这取决于周围地下水水质和任何受地下水影响的生态系统 这些价值的设置涉及咨询股东的进程，并且在实践中很可能关系到地下水的分配规划过程	问题 5：作为目标含水层的环境价值，是进行饮水供应，还是严格保护水生生态系统或生态价值？ 如果是——若问题 2 的答案是否的话(源水的水质不符合水质要求)，若用这些水进行补给，要进行地下水污染的高风险预计 如果否——地下水污染的低风险预计。然而，对低风险这不是一个充分条件
6. 地下水咸度和开采效率	如果自然条件下地下水的咸度高的话，开采前与源水的混合液中含盐量应当受到限制 在这种地方，密度可能也会影响流动，补给的淡水将在当地咸地下水之上形成一层透镜，使开采更加困难并且降低开采效率(如作为补给量的一部分，开采水的体积达到预期应用的环境价值)	问题 6：自然条件下地下水的盐度是否超过 10 000 mg/L？是否超过使用开采水的盐度标准？ 如果两者都是——实行低开采效率的高风险预计。含水层(尤其是含水层中地层)的水力特征，需要在第 2 阶段进行细心的检查 如果只有后者是——进行低开采效率的中等风险预计，但是对于低风险来说，这不是一个充分条件(例如，周围侧向径流速率快的咸含水层) 如果两者都否——实行低开采效率的低风险预计
7. 源水与含水层间的反应	对处于衰减带以外含水层的水、源水和含水层矿物之间的反应可能导致开采水水质的退化，或引起过分堵塞、含水层溶蚀 在第 2 阶段将进行全面的评价，但是在开始阶段反映潜在问题可能性的一些简单指标应当记录下来，将源水水质和当地地下水水质进行一定程度的对比	问题 7：氧化还原情况、pH 值、温度、营养物情况和地下水的离子强度与补给的源水是否相似？ 如果是——在源水与含水层之间进行有害反应的低风险预计，但是对于低风险来说这不是一个充分条件 如果否——很可能在源水与含水层之间进行有害反应的高风险预计，并且确保在第 2 阶段进行地球化学模拟
8.与现存地下水用户连接的生态系统和性质边界的靠近程度	附近现存的地下水用户和与地下水相连的生态系统很可能影响第 2 阶段所需的调查范围 典型的例子是，衰减带将含水层的滞留时间提高到一年 如果边界性质靠近 MAR 的地址，则衰减带可以延伸到相邻的含水层之下 由于 MAR，承压含水层中地下水的压力反应与水质反应相比可能传播相当长的距离	问题 8：在 MAR 地址附近(即 100 ~ 1 000 m)，是否有其他地下水用户，与地下水相连的生态系统或性质边界？ 如果是——可能对地下水用户或生态系统存在高风险影响，这在第 2 阶段需要特别注意 如果否——很可能对地下水用户或生态系统存在低风险影响，但是对于低风险这不是一个充分条件

续表 2.1

	评估所需的信息	问题及难度指标
9.含水层容量和地下水位	MAR 引起地下水水丘的隆起升高，依赖于含水层的水力特征，补给区的规模和补给速率 当含水层特性确认后，通常在第 2 阶段计算隆起。但是，过分的隆起会导致：涝、土层凸起、地下基本结构的洪涝、盐度潮湿、土壤盐渍化 因此，通常来说，对于大规模的补给工程，以位于浅水位地点的非承压含水层作为储存目标是不适合的 对于承压自流含水层，需要防止压力过大，并密封现存的井以防水流出	问题 9：在乡村水位埋深超过 4 m 或城市超过 8 m 的地区，含水层是承压的和非自流的还是非承压的？ 如果是承压含水层——进行水浸或过分的地下水隆起高度的低风险预计，但是对于低风险来说这既不是必要条件也不是充分条件 如果否(承压含水层)——进行水浸或过分的地下水隆起高度的高风险预计，但是第 2 阶段的调查可能显示这样的风险是可接受的
10. 潜水含水层中水质的保护	如果是潜水含水层，且准备开采后作为饮用水供应的，那么上面使用的土地和废水的处理(包括密集的园艺区和化粪池)应该小心处理或防止进入地下水的汇水区	问题 10：当开采水的使用包括饮用水供应时，含水层是否是非承压的？ 如果是——进行土地和废物管理中地下水污染的高风险预计 如果否——进行土地和废物管理中地下水污染的低风险预计
11. 裂隙含水层、岩溶或活性的含水层	如果含水层是裂隙岩层或喀斯特地貌，则需要提高开采储存水的能力，尤其是在周围地下水是咸水或水力梯度很大的情况下 由于补给区中补给水的快速运移，对于较大的衰减带也需要进行预备措施	问题 11：含水层是由裂隙岩层或岩溶介质，还是包含已知的活性矿物构成？ 如果是前者——进行补给水运移的高风险预计。扩大的衰减带是必要的。现存含水层的环境价值就会得到满足。在第 2 阶段要确保含水层基质的溶解和金属的游离 如果是后者——进行补给水运移的低风险预计。但是，对于低风险来说这不是一个充分条件
12. 成功项目的相似性	MAR 的一项基本原则是所有监测数据的确认和验证，应在公共领域完成，其中包括充足的运行数据，以便能够准确解释 这些信息对未来 MAR 项目在改善设计和运行、降低成本以及进一步完善指导方针方面是很有价值的 一个国家或州知识库中的这些数据对于倡导者是可以得到的	问题 12：在同一含水层中是否有另一个项目具有类似水源且成功地运行了至少 12 个月？ 如果有——若考虑当前项目的设计和第 2 阶段的调查及随后的风险评估，可以将现有项目确认和验证的数据考虑进去 如果没有——预计所有的不确定性因素需要在第 2 阶段的调查中解决
13.管理能力	在第 2 阶段调查的同时，对 MAR 运行不熟悉的倡导者需要获得适当的专业知识，以论证风险评估中剩余风险的低水平	问题 13：在 MAR 地点的运行方面是否有经验处理相同或更高难度的和水处理或涉及水质风险管理的供水调度？ 如果有——由于经营者有经验，实施水质差的低风险预计 如果没有——由于经营者缺乏经验，实施水质差的高风险预计。倡导者将从运行的此类系统中获得指导(如 MAR 运营商课程或含水层储存和开采过程的课程)或在进行第 2 阶段的同时雇用一个合适的管理者致力于有效的风险管理，以减少剩余风险

上述问题包括与含水层环境价值相关的源水的水质、开采水的预期用途、堵塞的可能性以及潜在的矿物质反应等。人们需要了解地下水水质与开采水使用的关系，以及作为饮用水供应或高保护的生态价值的地下水是否需要保护。人们还需查明附近是否有地下水用户或生态系统，含水层是承压的还是自流的，断裂是多孔的含水层的话是否存在同一含水层类似源水的项目，且支持者或其顾问、运行商在 MAR 或水质管理方面是否有经验。

MAR 的调查和试验的费用不是小事，完成这一程序，倡导者应当了解其拟议项目难度是高还是低，且在调查阶段掌握最有价值的各种类型的信息。因为在正常情况下首先确保至少获得 MAR 核心批准(见 2.1.2 部分)，这些调查的费用需在进行此类调查之前投资。

2.1.2 批准要求

在澳大利亚任何州或地域内，对于 MAR 项目的审批当前还没有统一的程序。对 MAR 项目审批可能需要进行的项目包括：

- 分享源水的权利，如雨水、再生水或其他来源，同时考虑环境流量和其他源水的用户；
- 建设调查井，ASR 井(含水层储存和回采)或开采井的许可证；
- 蓄水、覆盖岩土工程安全性、市容、昆虫和虫害滋扰及溺水危险的规划审批；
- 含水层环境价值的申报，说明周围地下水水质和目前的使用情况；
- 水补给到含水层要得到正式批准，以保护含水层的环境价值，防止水头变化过度，并保护人类健康和环境作为储存水开采后预期应用的结果；
- 分享含水层的存储空间的权利，我们认为储存量是有限的，并可能小于可收获的源水的体积；
- 开采含水层水的权利，可能作为累计补给量的一部分，这取决于该含水层过度开采的程度，并考虑其他的地下水用户和投资商，以免给他们造成不利影响；
- 转让水的权利，拥有将补给的权利转移到第三方的能力，服从水文地质的限制，不能选择压力水头已经很低的地点；
- 使用开采水的许可，以确保使用符合集水区管理计划和水质适合预期应用。

国家和(或)区域的水资源管理当局通常负责水量权利；环境保护机构通常负责水质和补给审批；当地政府负责规划和开发许可；健康委员会负责使用许可。若再生水是一个水源或进入饮用水管道或非饮用水供应管道被用来分配开采水，则水公用事业也将参与。

2.2 新台阶

假设入门级评估(第 1 阶段)显示该项目是可行的，难度并不足以阻止提议者，并且调整者没有发现其他障碍，则下一阶段进行源水、预处理方法和含水层的调查，以确定该项目是否确实能保护人类健康和环境，特别是含水层的调查。

第 1 阶段属快速定性评价，第 2 阶段则是基于已有信息和通过第 1 阶段预定的特定地点调查获得的补充资料进行的定量评价。信息必须在确认试验工程的运行是可行的条件下，项目才能按预期的设想开始运行。或试验工程建造后再进行全面的项目建设。项目开发的分阶段方法(见图 2.3 有助于避免浪费时间和金钱，可以通过将它定制到含水层中改进该项目的设计，它也允许将适当的投资用到 MAR 项目所涉及的替代或补充项目中。

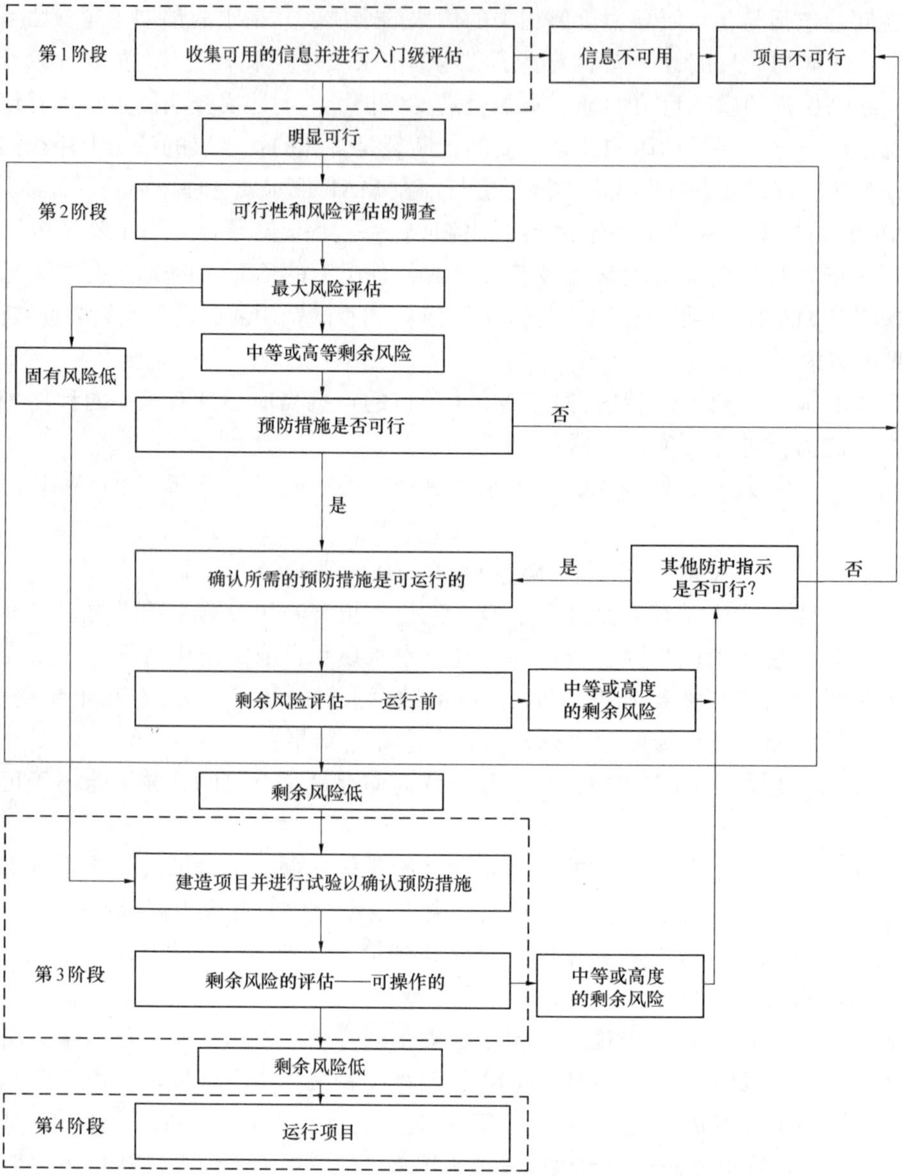

图 2.3　建立 MAR 项目的阶段以满足人类健康和环境的需要

(来源：Peter Newland，南澳大利亚州环保局)

第 2 阶段的调查能够评定风险并通过鉴定采取预防措施。这需要描述源水水质、基础设施和工程建议运行的信息，水文地质的性质表明：对于可管理的规划，所有的危害可通过足够的支持信息得到解决。如果需要进一步的资料，可能需要一个试验工程。

在这一点上成功的风险将得到更好的界定，并作出决策是否投资建设 MAR 项目或可选择的供水。这种决策将说明所有项目的成本和效益问题。

对于许多 MAR 项目在实施前不能得知当地的风险水平，通过适当的监测。众所周知的，如调试试验，第 3 阶段的监测为现行运作的 MAR 方案提供了依据，此外还为目前运作的项目提供管理计划的开发。第 4 阶段，需要进行持续的监测，以确保对人类健康和环境安全的风险能够得到控制。

2.3 健康和环境保护的风险评估框架

MAR 风险管理框架包括 12 项内容(见图 2.4)从属于四大类：

- 负责使用和管理循环水的义务；
- MAR 系统分析和管理，如风险评估和一系列的预防措施；
- 保障性要求，如员工培训，社区参与，研究和开发，社区参与，研究和开发，文件和报告；
- 审查，包括评估和审计程序。

需要执行所有 12 项内容以使得风险管理方法成功。

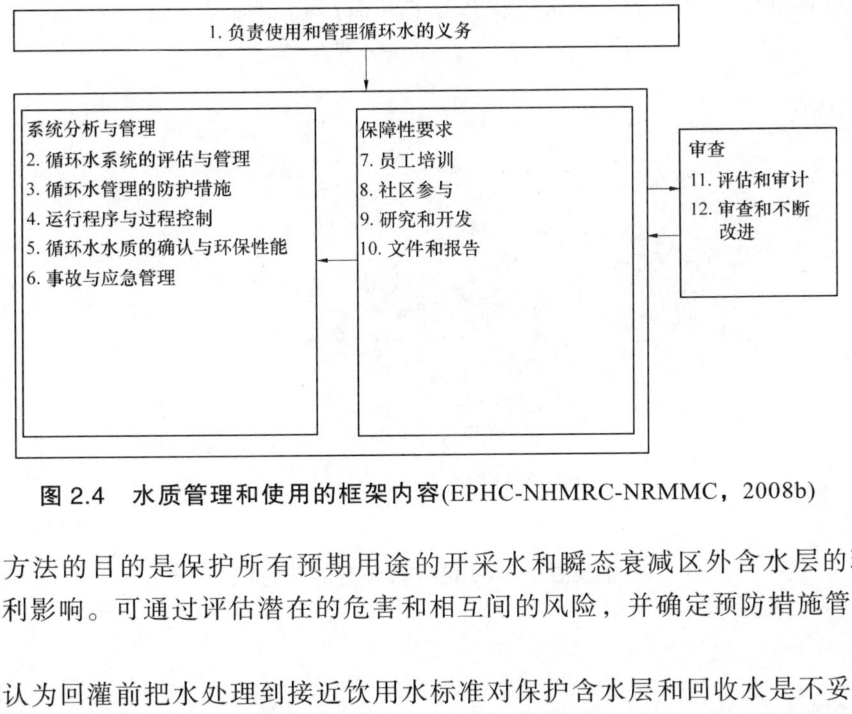

图 2.4 水质管理和使用的框架内容(EPHC-NHMRC-NRMMC，2008b)

这种方法的目的是保护所有预期用途的开采水和瞬态衰减区外含水层的环境价值以防止不利影响。可通过评估潜在的危害和相互间的风险，并确定预防措施管理风险完成。

简单认为回灌前把水处理到接近饮用水标准对保护含水层和回收水是不妥当的。例如加氯消毒以持续消除含水层中的病原体，可导致一些含水层开采后水中氯仿过量。在一些地方，将饮用水注入到作为饮用水的含水层后，由于其与含有砷的黄铁矿间的反应，会造成开采后砷浓度过高。源水淡化成高纯度水后，相比处理程度较低的水，容易使含水层溶解更多的矿物。

因此，介绍一种考虑含水层与补给水相互作用的 3 个科学方法：

(1)可接受的危险消除——水在衰减带内通过土壤和含水层廊道的过程中允许衰减(见图 2.5)，(如病原体灭活，一些微量有机物的生物降解，有限数量的养分吸收)。

(2)低效率的危险消除——这些危害需要在补给之前消除，因为他们不是不能消除(如盐度)就是消除是不连续的(如任何金属的吸附和不是随后降解的有机物，或过多的营养物或悬浮固体)。

(3)含水层的相互作用造成的新危险(如金属运移，硫化氢，盐度，可靠性，硬度，或放射性核素)。有必要改善补给水的水质以避免这些危险(如改变酸碱度、减少氧化状态或减少营养物)。

含水层对任何水质危害的反应取决于含水层特定的条件，包括温度、氧气存在与否、硝酸盐、有机碳和其他营养成分与矿物质，及事先接触到的危险。

水质受到 MAR 影响的含水层范围可能比衰减区大，但边界外区域的水质需要保持一直满足含水层初始的环境价值，MAR 运行对水头(压力)的影响范围更大，尤其是在承压含水层，并可能延长数公里。作为开采水使用，如果含水层初始含盐量较高的话，那么就需要确定一个储存带及开采的时间，以便使开采水按预期的用途使用(见图 2.5)。

图 2.5 中虚线为衰减带的外部边界。表示 MAR 补给区与周围地下水水质未受影响监测井之间的最大间距。由于衰减带只界定了可接受的衰减，随着 MAR 运行的停止，衰减带将缩小和消失，最终整个含水层将达到其所有最初的环境价值。

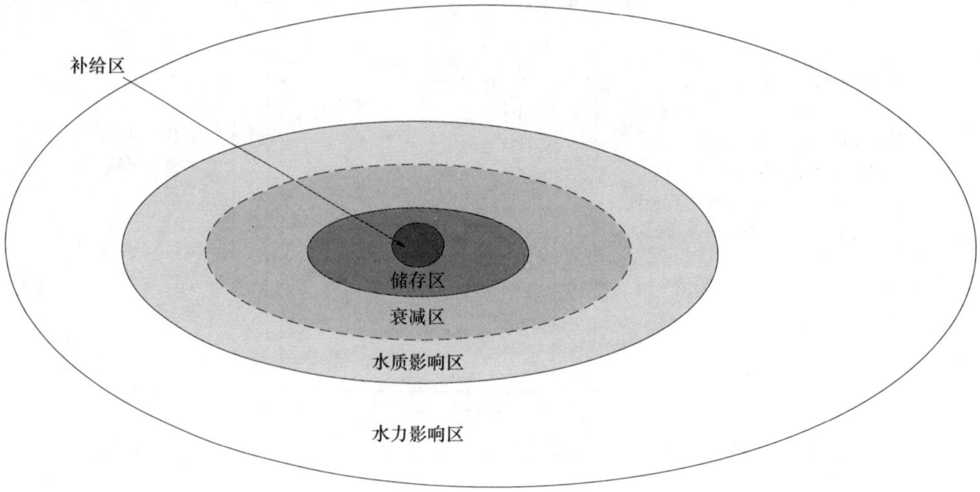

图 2.5　MAR 运行影响区的示意图

3 水力因素

3.1 堵塞

堵塞可导致多孔介质渗透性的减小。它是可管理的含水层补给实施问题中最为严重的问题之一，并在一定程度上影响着大多数的实施操作工作。堵塞是井灌系统中一个较显著的问题，因为水进入含水层中的速度比进入回灌池要快几个数量级。依此类推，回灌池中水入渗的速度比自然补给情况下要快几个数量级。

堵塞是在补给的过程中产生的，它是源水、水的成分以及多孔介质共同作用的结果。这种现象主要发生在注入补给水的区域周围、入渗系统的表面，或集中发生在井灌系统中的滤水管、砂砾或相邻的含水层附近。

堵塞导致流速下降，从而制约蓄水量，并限制了可管理的含水层补给运行的规模和经济可行性，而且在严重的情形下，可导致可管理的含水层补给的失败。如设计不当、管理不善，可管理含水层补给系统中回灌井可能在数天或数周内发生堵塞，而回灌池可能在数周或数月内发生堵塞。

堵塞是由一个或多个物理、化学或生物过程共同作用的结果，例如：

- 悬浮物的渗入；
- 微生物的繁殖；
- 地球化学反应；
- 滞留空气和气体。

(1)悬浮物的渗入。由补给水中悬浮物的渗入引起的土壤或含水层空隙的堵塞会导致低渗透性的堵塞层的形成。堵塞现象在固体悬浮物浓度很高的地方一般是非常严重的，但是也有其他的影响因素。其中，较重要的是水的流速，它使得累积的沉淀物在多孔介质内沉淀下来；其次是多孔介质的水文性质，它决定着系统内颗粒沉淀物的范围和渗透性。

(2)微生物的繁殖。含水层中微生物的堵塞以微生物细胞繁殖和产生细胞外多聚糖(生物细胞膜)的形式出现。当补给含水层中含有的养料(如有机碳、氮、磷)刺激微生物的活动时，细菌就会产生细胞膜。由池底水藻的生长引起的堵塞是地表下渗系统中的一个难题。它可能会与碳酸盐矿物质发生化合作用产生沉淀，例如方解石就是随着光合作用中二氧化碳的消耗而产生的。水温的日常或季节的变化影响着水藻的生长和补给水的黏性，并最终决定着下渗的速率。其他因素的影响作用都相差不大，较为典型的是，夏季的下渗速率比冬季要高。

(3)地球化学反应。与地下水或含水层物质不平衡的水分补给可以引起化学反应。它们会产生难溶的沉淀物，改变多孔介质的渗透性。这些化学反应可能包括：

- 分解；
- 沉淀；

- 离子交换；
- 离子吸附；
- 氧化–还原。

如果含水层富含离子或锰的矿物，那么补给水中所含的氧或硝酸盐会刺激微生物的活动，使离子或锰的氧化物及氢氧化物沉淀下来，从而导致含水层堵塞。在补给水和含水层中的黏土之间进行的阳离子交换会引起涨水或弥散现象。这种现象在有水敏性黏土(如高岭土)的地方和补给水的盐度比周围地下水低的地方非常普遍。另外一种地球化学反应——分解，可能会通过增加渗透性(如在方解石溶解的地方)而不产生堵塞现象。已知的实例表明，这种情况会引起井的不稳定。离子的氧化物和氢氧化物以及碳酸钙的沉淀是地球化学堵塞的主要形式，但是它们并没有被广泛地认为是主要的堵塞机制。这是因为它们与其他形式的堵塞一样，或是因为它们需要经过很长的时间才会形成。许多地球化学反应，尤其是氧化还原反应，都是通过微生物催化的；这样，在许多情形下，会难于区分化学和生物过程。

(4)空气滞留以及气体阻塞。如果水是直接泄落到井内的，就会产生气泡并阻塞孔隙，阻滞水流，导致堵塞。溶解的气体由于温度变化也会从水中释放出来，比如低温的补给水与温暖的地下水相遇时。或者生物地球化学反应的结果也会导致气体堵塞，例如，反硝化作用能够产生氮气。

关于可管理含水层补给系统中堵塞问题的更多信息可参阅 Olsthoorn (1982), Pyne (1995), Pérez-Paricio 与 Carrera (1999), Bouwer (2002) 与 Pavelic 等(2007)的相关研究。

3.1.1 堵塞治理

堵塞治理的一般方法是降低补给速率，增加压力。用来判断和解决上述四种主要形式的堵塞的方法可见表 3.1。表中指明了通过何种方式可以减少或防止堵塞：

- 水源充分的预处理；
- 场地的适当的选择；
- 设计良好的装置的建设；
- 执行正确的操作控制、改造措施和监测措施。

有多种方法可以防止或消除堵塞，在一定程度上它们比治理堵塞更为有效。例如，去除源水中的固体悬浮物、有机物质以及营养物质的预处理技术能减少堵塞，当然这也需要一定程度的改造工作。一些预处理技术在处理阻塞方面是很有效的，但是有较大的负面效应。例如，源水的消毒会减少微生物活动，但是同时会在含水层中产生消毒后的副产品。在井灌的过程中杜绝掺气的措施相对比较直接，只需要在静态水位以下放入入水管，并安装一个水阀，以保证持续的正压力。

改造的目的是通过复原堵塞区域的水力性质(也就是清除堵塞)，使回灌井和回灌池的性能回到初始的状态。对于回灌池而言，清除堵塞是相对直接的方法，因为它只是进行简单的定期疏干处理和偶尔的清除堵塞物质。

例如，在土壤含水层处理系统(土壤渗滤系统)中，渗滤期要持续数天至数周，渗池也应被疏干并干燥几乎相同的时间，当然这可能需要季节性的调整。

有多种机械方法曾被用来恢复注水井。这些方法包括反冲洗泵(它的速率比补给速率高)，水击(变速泵)或高压喷射(也被称为气举)。恢复是周期性进行的，通常是每天，这取决于补给率下降的快慢。通常情况下，只有少数百分之几的补给量能预知。

表 3.1 阻塞的类型、原因、判断以及处理方法

堵塞机理	原因	减少和防止	改造措施	方法技术
固体悬浮物的过滤	1．源水中的微粒或管道中的沉淀 2．水处理副产品(如明矾) 3．池中藻类的生长	1．提高源水的预处理，尤其是悬浮固体的去除 2．避光(藻类生长) 3．经常回洗灌井，或疏干渗池	1．抽水 2．水击 3．喷射 4．干燥和刮洗池底	1．源水、土壤、含水层中总固体悬浮体、浊度，以及粒子大小的分布 2．旁路过滤测试 3．单位出水量变化(井)
微生物的繁殖	源水中含丰富的基质(尤其是 N、P 和 AOC)	1．提高源水的预处理，尤其是营养物质或氯处理 2．经常回洗 3．改变池干湿的周期	1．抽水 2．氯冲击处理剂量 3．酸处理 4．加热 5．用钢丝刷清理 6．多酶溶液 7．池底的干燥和清理	1．水质分析 2．摄影(渗池)和井下钻孔电视检测 3．土壤和含水层样品的化学微生物分析
化学沉淀，离子交换	1．氧化还原反应 2．源水、地下水和含水层矿物的不相容 3．水敏性黏土的存在(膨胀/分散)	1．提高预处理 2．调节源水的 pH 值和阳离子的浓度 3．降低源水中的 DO 4．经常清洗	1．抽水 2．用钢丝刷清理 3．酸处理 4．水池的清理	1．水质分析(尤其是氧化还原条件和矿物溶解度) 2．电镜分析回洗水或土壤和含水层样本(深井刮壁) 3．摄影(渗池)和井下钻孔电视检测
掺气，气体堵塞	1．自由落水 2．源水中气体的存在 3．生物化学反应	1．避免泻落 2．脱气器 3．源水的除气 4．预防反应的产生(如氮气的产生)	1．抽水 2．等待气体的再溶解 3．改善井的设计	分析水温和特殊气体

注：①AOC 为可同化的有机碳；DO 为溶解氧。
②资料来源：根据 Perez-Paricio 和 Carrera 1999,Pyne 1995 改写。

已用的化学方法包括添加氧化剂(如氯)、凝聚剂(如氯化钙)，或者在严重堵塞的情况下用酸。改造措施作为清除渗井和渗池的惯用操作方法是十分有效的，这已被一些场合持续数十年的实施操作现实所证明。

3.1.2 预测堵塞的工具

已经有大量的技术用来预测堵塞。技术复杂性的范围从简单的水质指标的测算到复杂的过程模拟。主要技术概要如下。

3.1.2.1 补给水水质参数方法

补给水水质参数方法对可管理的含水层系统的潜在堵塞可能是一个很好的指示。有用的补给水的水质参数包括：

(1)物理堵塞：浊度、总固体悬浮物、膜过滤指标。

(2)微生物堵塞：总有机碳、溶解有机碳、可同化有机碳。

当多种形式的堵塞同时发生时，单一参数的应用会产生很多问题。预期的源水或目

标含水层堵塞潜力的特征描述应当包括源水中物理、化学以及微生物组成的评价。评价需要考虑时间变异性。土壤和含水层的生物化学性能、矿物学特性以及周围地下水水质也应当有所确定。

3.1.2.2　实验室土柱研究

利用土壤和含水层芯料以及源水进行的实验室土柱研究能够判断堵塞的范围，及源水预处理的必要程度(例如，Rinck-Pfeiffer 等，2000；Wood 等，2005)。然而，由于地下特征的空间差异，小范围的测试仅能提供一种指导，并不常用于精确地预测整个场地范围。

3.1.2.3　图解法

用来预测堵塞的简单图解法已经给出。Hutchinson 和 Randall(1995)应用含水层的导水系数和 MFI 数据提出了广泛预测注水井的物理堵塞速率的典型曲线。Taylor 和 Jaffé (1990)应用实验室土柱研究的实验数据，从多孔介质内生物数量水平上，测算了渗透系数的相对减少。

3.1.2.4　数量或质量平衡模拟

复杂的数量模拟程序将流动、运输以及进行孔隙度和渗透性变化的特殊场地预测的生物地球化学反应耦合在一起，这方面已经得到了发展和广泛的应用(Pérez-Paricio 和 Carrera, 1998)。但是，这些编码相当复杂，数据要求非常精确，而且只适合当前的研究。或者可以进行简单的质量平衡模拟来估量物理、化学和微生物过程中对渗透性的可能影响程度(Pavelic 等，2007)。

3.1.3　实施可持续的 MAR 对补给水水质的要求

几十年的实施运行历程已经表明，预防堵塞是一种比改善阻塞更好的选择，尤其是对于井灌系统来说。同时，多种多样改造措施的存在使其可以得到很好的结果(见表 3.1)，并保证回灌水达到适当的水质要求，因为充分的预处理是保证成功的、可持续的、长期的可管理的含水层补给实施的关键因素。虽然各种形式的堵塞可以通过预处理得以避免，但是在实际中，在与预处理相关的花费和可接受的堵塞程度之间存在着一种平衡，而且可接受的堵塞程度取决于改造处理的类型和频率。澳大利亚可持续的可管理的含水层补给实施中要求水质的确定指标值是人们极力追求的目标。但是到目前为止，这个目标仍尚未达到，还是一个需要大家积极、努力研究的领域。

荷兰可管理的含水层补给的实践者们通过实验总结已经测定，向沙质的含水层中注入的源水的适宜性，对于物理堵塞来讲要求膜的过滤指标小于 3 ~ 5 s/L^2，对于生物堵塞来讲要求可同化有机碳小于 10 μg/L(Olsthoorn 1982; Hijnen 和 van der Kooij，1992)。但是，这种单一的指标值对于澳大利亚而言是不适用的，因为澳大利亚源水的类型和地质条件比荷兰有更大的可变性。Pérez-Paricio 和 Carrera (1999)开发了源水水质与注水井和地表渗池的可能堵塞程度相关的指南。这些指南提供了一个有用的起点，但是缺点是没有考虑到土壤和含水层条件。

根据源水水质和含水层类型，对 14 个含水层存储和开采场地的堵塞速率进行分类。结果表明这两种因素都可能影响堵塞速率(见表 3.2)。对于在冲积含水层中的含水层存储和回采项目而言，总固体悬浮物和总有机碳低于 10 mg/L 的源水的堵塞速率比更浑浊或营养物质更丰富的水源要低。对于水质较差的水源，石灰岩含水层的堵塞速度最慢，其次是冲积含水层，裂隙岩含水层的堵塞速度最快。这种结果并不足以令人感到奇怪，因为石灰岩含水层的堵塞物质可以通过基质溶解的形式方便地清除，而且冲积(沙和砾石)含水层可以提供比裂隙岩含水层更多的水流路径。为了满足操作要求，判定堵塞潜力和判别适合于含水层的水质目标的特殊场地，需要在所有的场地中进行评价(Pavelic 等，2007)。

表 3.2 在 14 个含水层储存和回采场地中由含水层类型和源水水质决定的堵塞速率

含水层类型	不同源水水质所确定的堵塞速率		
	低 (TSS >10 mg/L； TOC >10 mg/L)	中 (TSS 为 1~10 mg/L； TOC 为 1~10 mg/L)	高 (TSS <1 mg/L； TOC <1 mg/L)
石灰岩	低→中	低→中	无可用数据
沙和砾石	中→严重	低→中	低→中
裂隙岩	严重	无可用数据	无可用数据

注：TSS 为总悬浮固体；TOC 为总有机碳。

3.2 开采水的咸度——开采效率

可管理的含水层补给实施技术的给出依赖于储存可利用水的空间，以及提供可开采一定数量优质水的能力。当水下渗或被注入到含水层中时，这些水会与周围的地下水混合，引起孔隙水速率(弥散度)以及溶解物质的浓度梯度的局部变化。混合现象也会在含盐的含水层中发生，原因是密度的差异会引起对流运动。不同水体之间的混合程度依赖于局部含水层的水力特征以及周围地下水的水质。

限制混合对于微咸水或咸水含水层中淡水的储存是非常重要的。当周围的地下水对于开采水而言盐度过高(或包含其他的浓度过量的物质)时，混合必须加以限制，以使满足水质要求的开采水的体积最大化。当源水和周围的水体水质很好时，混合就显得不太重要了，除非地表以下的地球化学过程产生的反应副产品(如离子，砷)超标。

本节需要解答的主要问题为：

- 可管理的含水层补给实施者可以开采多少补给水？
- 补给水和盐度较高的周围地下水的混合程度是多大？
- 适合利用的开采水的比例有多少？
- 含水层中补给水的分布情况怎样？它们能运移多远？
- 补给水能够储存吗？它会不会被其他的地下水用户抽取？

3.2.1 开采效率的定义

开采效率指的是能作为适合预期使用的开采水占注入或下渗的体积百分比(Pyne，1995)。这样，开采效率为 1，表明开采水的体积等于补给水的体积；开采效率为 0，表明抽出来的补给水都达不到适合的水质标准。

实际上，按照预期使用的要求，开采过程受到开采水中固体溶解物的最大允许浓度的限制。例如，作为饮用水使用，源水要求总矿化度的浓度上限是 500 mg/L，以达到澳大利亚饮用水的标准。灌溉供水主要依赖于土地利用类型的耐盐度。在开采水与淡水混合的实例中，允许的总溶解盐浓度的升高，会引起开采效率的增大。

有许多物理、化学和微生物物种可用来监测地表以下的补给水。其中许多在自然条件下就存在于源水之中的，其他的可以人为的添加。一些常用的监测补给水的对象包括：

- 总矿化度；
- 电导率；
- 氯；
- 溴化物；
- 氟化物；
- 温度；
- 氘；
- 氘氧–18。

活跃的化学物质，以及用来缓冲系统的物质，其性质不稳定，不能被用来作为监测对象。

许多因素影响着含水层中补给水的运移和混合，并最终影响着开采效率。这些因素可以分为两类：水文地质变量和管理变量。

水文地质变量是一个场地所固有的，包括：

- 含水层的厚度、导水系数、孔隙率以及非均匀性(弥散度和扩散率)；
- 区域和局部的已经存在的水力坡度；
- 周围地下水的水质和密度；
- 补给水和周围地下水的溶质浓度；
- 补给水和周围地下水的密度差异。

管理变量包括：

- 井的设计；
- 开采井；
- 适用地区的注水井；
- 总存储量；
- 注水井和开采井的间隔；
- 下渗和注水速率及流量；
- 开采速率及流量；
- 开采井的位置；
- 在含水层中的滞留时间；
- 使用周期。

水文地质变量并不是在任何场地都能得到控制的，但是在场址选择过程中应该把它们考虑在内。管理变量在不同程度上可以得到控制，例如通过调整储存水时间(井的注水模式)或回灌池和抽水井之间的距离引起的地下水流速的变化，可以调整滞留时间。流域系统中未饱和区域的作用也需要加以考虑,例如关于储存在土壤剖面中的盐的运移潜力。

3.2.2 澳大利亚 ASR 实施项目中的开采效率资料

开采效率数据是从澳大利亚在过去的 10 年里的几个试验和运行的含水层存储和开采场地中得到的。澳大利亚 6 个场地的资料(根据共同的阈值浓度 1 500 mg/L，对浓度和体积数据重新计算)表明：

- 由于上述因素，不同场地间的开采效率也明显不同(从小于 0.02 到大于 1.0);
- 在没有主要的次生孔隙度特征的石灰岩含水层中，中等到高的开采效率可以实现；
- 由于未开采的补给水的盐度缓冲(见图 3.1), 在含水层的储存和开采工作的连续循环过程中，可以提高开采效率；
- 含盐量高的含水层的特征是开采效率低，其形成原因是可以接受的混合程度低，能够产生补给水浮力分层的密度效应。

目前无其他形式的可管理的含水层补给的类似资料(如在沿海的沙质含水层中地表补给的方法)。

3.2.3 混合的管理以及开采效率

从国内和国际的经验中获得了关于场地的选择和使开采效率最大化的操作管理的指导原则，包括以下方面：

- 避免地下水的咸度过高。
- 降低源水的咸度。

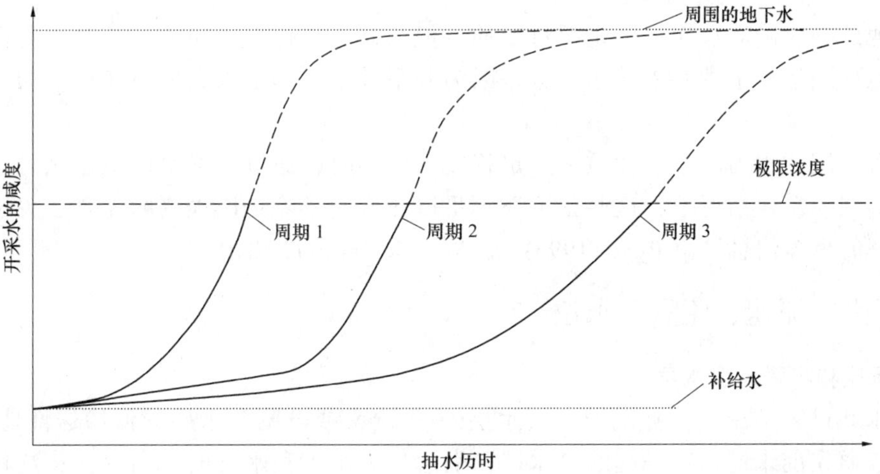

图 3.1 当达到最大允许浓度时，开采阶段中抽取的水中盐度的变化以及抽水的停止时刻。未开采的补给水形成了一个含盐的缓冲带，并在随后的循环周期中开采效率逐步提高。如果在达到极限浓度之后还允许抽水的话，则虚线表明了盐度的变化

- 选择周围地下水流速较低的含水层。
- 避免高度不均匀的含水层。
- 避免渗透性高的含水层，尤其当周围地下水咸度较高时。

- 保证目标含水层(以及适用的未饱和区域)：
  ——有适当特征的；
  ——有适当的储存间隔；
  ——有设计合理的注水/开采井。
- 保证可储存充分体积的水(具体情况视尺寸大小而定)。
- 选择开采井的位置，以最大程度地截获补给水。
- 减少存储时间来限制补给水的区域性的运移。

其中的一些原则可能与可管理的含水层补给的目标相冲突，这就要求做出一些权衡。例如，为避免补给水低开采性，不要使用高渗透性的含水层或子含水层；与此形成对比——减少井堵塞的原则在前面已有所概述。

可管理的含水层补给系统的设计方案为满足水质目标提供了一些弹性的标准。例如，在地表下渗后，用于抽水的开采井的位置就能够确定下来，因此他们能大量地用于抽取补给水，减少补给水的延伸超过蓄水区域。在不需要与周围地下水混合的情况下，这是非常有用的。或者井的位置可以选择在能持续地抽取补给水和周围地下水混合物的地方。

通过开采水与可供选择的微咸水源相混合的方式或者通过提供反渗透的后处理方法，可以提高开采效率。

3.2.4 开采效率的评价

开采效率受到很多不确定性的相互作用的变量因素的制约，因此很难在可管理的含水层补给进行之前预测开采效率。有两种主要的方法可以用来评价开采效率：场地试验和地下水的模拟(运用分析或数字技术)。

场地试验对确立操作程序是很重要的。但是，当进行小规模的试验时，它们经常不能可靠地预测运行中的开采效率。这是因为开采效率依赖于水文地质变量，如含水层的弥散度。

当含水层的水力和运移性质在一定程度上可知时，通过地下水模拟预测可以测得开采效率。在大多数情况下，这些预测还不足以让人们排除场地预测的需要。关于注水井开采效率的更多信息已由 Pyne (1995) 和 Pavelic 等(2002)给出。

3.3 压力、流速、流量和水位

3.3.1 承压和半承压含水层

注水井过大的压力能够割裂弱透水层(一种定义或分割含水层的低渗透性地层)。此举可以连通先前未连通在一起的含水层，引起压力水平和水质的逆向变化；这种结果能使实施含水层存储和开采获得更多的利益。

过多的压力也会引起质量较差的注水井和其他的井产生问题，使水流到其他的含水层或地表。承压含水层的加压也会使此含水层中存在的完整井成为自流井。从这些井中排出的水可能存在需要解决的困难或危险。

在开采储存水的过程中，含水层的测压管水头会降低。这就需要更多的能量来抽取受水力因素影响的钻井中的水，以减少它们的出水量。在一些情况下，尽管含水层中有额外的储存水，由于含水层的储存和开采运转，现有井的所有者也可能会处于不利地位。

过度的减压可能使可压缩的弱透水层和其他上面的地层，或未饱和的承压含水层被压实。其中的任何一种情况都会引起地面沉降。如果承压层之下的含水层经常遭受化学或物理的侵蚀，过度的减压最终也会导致注水井或弱透水层出现问题。

承压含水层和半承压含水层中与压力、流速和流量相关的危险可以进行控制管理，包括以下预防措施，每一项措施的讨论如下：

- 注水压力的选取；
- 估计可能的自流区域；
- 含水层的组成要素以及流体压力转移的确定；
- 适当的时段内，补给量和开采速率的限制；
- 限制塑性变形；
- 最大注水压力的计算。

3.3.1.1 注水压力

应该正确选定注水压力以保证它不超过对弱透水层底部的干覆盖层的压力。在保守的情况下，压力(P)可以从 $P<15\,d$ (kPa)开始估算，这里 d 表示从地表到覆盖在含水层上的弱透水层底部的距离，单位是 m，并假定干重的密度大于 15 kN/m^3。

最大的压力不应受到阀门的限制，因为阀门毕竟可以做一些调节。特大型水泵会产生不可挽回的损失。正常情况下，注水压力更易受到能效和避免井周围堵塞层压实需求的限制。

在特殊情况下，有需要进行最大注水压力的更精确计算，因此需要进行以下内容：

- 测量弱透水层及其上覆物质(覆盖层)的容积密度；
- 测量覆盖层中测压管液面高程；
- 进行岩土测试以估算弱透水层的强度；
- 通过岩土评定计算爆破压力，来说明弱透水层的弹性性质。

3.3.1.2 自流区域

应当通过估算的注水速率和从抽水试验中得到的含水层水力性质来估计容易形成自流的区域。在所有者允许的情况下，应该鉴别一下这个区域中穿透目标含水层的所有井，而且预测所有能产生溢流的井都应该密封。如果不可行的话，就有必要限制补给速率，以使最近的未封闭的井不自流。

3.3.1.3 水压和组分的转移

校核受影响井的建筑记录以确定含水层中流体的压力和组分的转移是否是可接受的，以及井的回填是否是必要的。

3.3.1.4 体积、补给速率和开采速率

限制流量和补给速率及开采速率使之适合当地条件和含水层容量的周期(如年、月、日)是非常重要的。正常情况下，降低承压含水层中的水头值(使含水层非承压)是不可取的，因为这会导致压实和不良的地球化学变化，如还原金属矿物的氧化。同样，抽水泵也应该安装在含水层顶部水位之上。

3.3.1.5 塑性变形

如果弱透水层是不固结的，在抽水期间，随着水头降低对发生塑性变形应进行定量评价，并通过限制抽水速率，使变形达到可接受的水平。

3.3.2 潜水含水层

潜水含水层的补给可以增加储存量，保护受重压的含水层中依赖地下水的生态系统。但是，如果地下水位升的太高，非承压含水层的补给也可能会产生不利影响。例如：

- 水涝；
- 地下室、地下电缆管线和洼地的淹没；
- 减少交通路线、坑槽，增加道路养护的花费；
- 植物的缺氧效应；
- 增加地下水排入下水道系统；
- 增加建筑基础的不均匀冻涨效应，对建筑物造成破坏。

关于开采蓄水，降低地下水位可能会增加其他地下水用户的抽水花费，还会减少浅井的排泄，也可能调动机器，在最需要水的时刻，减少地下水排泄到依赖的生态系统的数量。

含水层中压力、水流或与体积相关的危险管理控制包括以下预防措施：

- 安装溢流转换装置以防止补给水头过大。
- 杜绝使用注水泵。
- 避免选择如下渗透场地：
  ——在斜坡上；
  ——在地下水位浅或不透水层以下；
  ——濒临地表洼地或依赖地下水的生态系统。
- 根据当地的条件，定期地限制补给量和补给速率(例如按年、月、日)。
- 在渗透期间，增加补给池或渗渠的干燥时间，以降低地下水水丘的高度。

在任何时间地点，当补给开始后，矩形渗池下面地下水壅升的高度可以通过一个等式计算出来(Hantush，1967；Bouwer，1978，2002)。计算结果依赖于渗透速率、含水层透水率和孔隙率，以及渗池的大小。

图 3.2 显示了位于潜水层的初始饱和含水层厚度为 10 m，水力传导率为 10 m/d，有效孔隙率为 0.1，面积为 100 m×100 m 的渗池中央水位升高的高度。土壤的类型和源水的水质会通过堵塞而影响补给的速率，从而也会影响地下水丘隆起的高度。例如，如果隆起的高度限制在 4 m 或者 8 m，则相应的补给速率需要小 30 m/a 或小于 80 m/a。

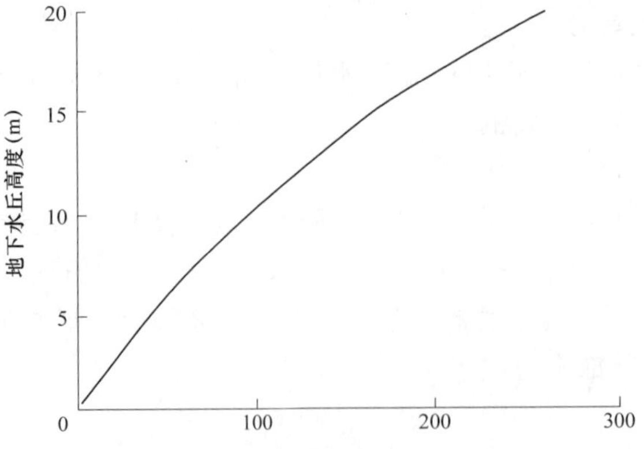

图 3.2 对于不同的补给速率，经过一年的补给后，渗池中央地下水可能隆起的高度

3.4 裂隙岩体含水层和岩溶含水层中污染物的迁移

在裂隙岩体含水层中，大部分水通过岩石中的裂隙流动，因此水只与岩石体的一小部分接触。裂隙岩体含水层的有效孔隙度要比那些水流主要在相互连通的多孔介质的孔洞中流动的含水层小得多。

水在相互连接的裂隙中流动，也可能在含水层中以三维形式曲曲折折地流动，因此从蓄水层或开采井注入裂隙含水层中的水要比在多孔含水层相同体积的水流得远(见图 3.3)，补给水的分布特征(也即衰减地带的大小)也因此比较难预测。如果含水层起初是微咸水，则开采效率也会随之降低。含水层抽水试验通常能揭示低的有效孔隙度；但是，如果结果很难预料，则当评估所要求的衰减地带的大小时，就要用到有效孔隙度的最小实时估算。

在岩溶含水层中，水主要在具有溶解特性的岩层中流动，例如在已经扩大的断裂、裂缝、流管和洞穴中流动。水的最小流动阻力或优先流动行程的尺度是几米。岩溶含水层通常水也允许在内部的颗粒间流动；整个含水层也可以传导水，但水的流动主要发生在具有流动特性的岩层中。如在岩溶含水层的双重孔隙系统中，水要经过更长的时间来代替原来的来自多孔介质的地下水。

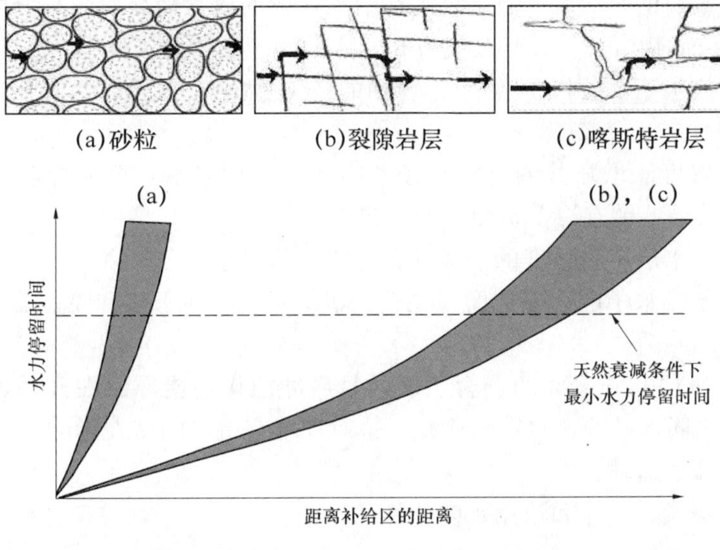

图 3.3 补给水在具有均匀多孔介质(a)、裂隙(b)和岩溶特性(c)的含水层中迁移的情况

(补给水在裂隙和岩溶含水层中迁移得比较远。例如，病原体衰减需要的停留时间，裂隙和岩溶含水层所要求的衰减地带会比较大，并且很难定义)

关于在优先流道的流动最快的含水层基质的表面积也因此要比流过含水层基质的表面积小,所以含水层的生物膜与流动快的补给水的成分发生化学反应的机会就会减少。大多数生物降解速率就会受表面积的限制，因此比流过含水层慢。水中的某些成分在生物降解达到符合周围地下水环境值的要求前潜在的运移很远。岩溶含水层的衰减地带因此会比预测的无岩溶特性的含水层大。

这种情况并不意味着避免使用岩溶含水层用于可管理的含水层补给。然而，它表明建议者需承担很大的责任来证明可管理的含水层补给方案不会对其他水的供给或连着的水体产生消极的影响。单独考虑具有水利特性的含水层很难充分地评定衰减地带的大小，但是来自附近的可管理含水层的补给方案的历史信息的指导在评价对人类和环境健康可能的危害时具有重要的价值。

当衰减地带依赖于病原体的失活或营养物和污染物的生物分解时，它们的范围不应该比通常所接受的校准者的范围大。裂隙含水层和岩溶含水层要求多于一个可信的观测孔以确认衰减带以外的浓度符合含水层的环境值。

3.5 含水层的溶解以及井或弱透水层的稳定性

补给水可能与含水层的基质发生化学反应，导致矿物的溶解或含水层的体积或强度的下降。

导致含水层溶解的原因包括：

- 酸性的补给水溶解碳化盐矿物(如方解石和白云石)；
- 补给水中的有机碳在含水层中氧化产生酸性环境，这也会溶解碳酸盐矿物；
- 含氧的补给水氧化硫化物；
- 与矿物不平衡的补给水，通过淡水补给水在咸的或含盐介质中与分散的黏土矿物发生化学反应导致矿物溶解，并且改变有孔介质的物理结构和强度；
- 阳离子交换吸附(特别是补给水是含钠的)分解黏土矿物和改变有孔介质的物理结构。

含水层的溶解可能增加井的有效直径，因此会增加产水量并能抑制慢性的堵塞问题。然而，含水层的溶解有许多负面影响，包括：

- 由于周围含水层矿物强度的下降导致外露井的坍塌；
- 产生浑浊水或水中的沙子能磨损泵轮、堵塞灌溉的沥干架和喷雾器，并且在配水系统中产生损害；
- 临近的蓄水层和开采井的弱含水层的最后冲蚀作用能导致在开采过程中弱透水层的倒塌和与有不同水质的含水层的连接，这对环境有持久的潜在影响；
- 有利于金属的运移；
- 开采水的味觉、气味和饮用水安全性的反面影响，特别是在硫化矿物氧化和衰减之后；
- 含水层基质中运动的黏土颗粒会引起堵塞，可导致注入井的比容量缓慢减少。

3.6 与其他包括依赖地下水的生态系统的地下水用户的相互作用

可管理含水层补给工程的管理者需要考虑工程对当地地下水用户的影响，以及用户对工程的实施所产生的影响。其他的可管理含水层补给工程的管理者也会造成某些特定的问题，这些都应在流域及地下水的管理计划中给出以避免未来潜在的冲突。那些位置近的可管理含水层补给工程可能导致对源水或含水层储存空间的竞争。

最近回灌工作者倾向于：有足够的地表水时进行回灌，地表水不足时进行开采。因此，对含水层的压力水头有双重影响。例如，在含水层储存和开采的情况下注入水的过程中每一次操作都会增加其他含水层的水头梯度，并能相应地增加要抽取的开采水的水头。因此，每一次操作都能增加费用以及下一次操作时所需的压力。

地下水依赖的生态系统(如泉、溪、沼泽地、湖泊、地下水湿生植物)潜在地受到地下水水质和水位线的影响，也受到超过正常季节性的和每年的变化范围的地下水排泄速度的影响。因此，可管理含水层补给也会影响生态系统。例如，通过依赖于地下水的沼泽地持续供水。依据生态系统管理的客观性可能产生好的和不好的影响。保存生物的多样性(包括迁徙的鸟)是Ramsar湿地首先考虑的(Ramsar Convention是关于保存和持续使用湿地的国际性条约)，当评价临近的可管理含水层回灌工程时，保存生物的多样性必须考虑。

4　水质因素

应用于MAR水质管理的系统分析和管理框架部分(元素2~6)显示了水质的一些风险(见表4.1)。对人类健康或环境的危害可能来源于：

- 回灌的水源；
- 天然地下水；
- 含水层矿物与补给水的相互作用；
- 处理过程和保养程序中的副产物。

危害事件如断裂岩层中污染物的迁移，或对其他地下水用户的影响，在“水力因素”部分已讨论过。

表4.1　MAR工程水源、地下水和含水层矿物中关键水质危害总结，及特殊危害例子和预防措施

险事	来源	例子	预防措施
病原体	S，(G)	病毒	足够的含水层停留时间
无机物	G，A，S	砷	在回灌过程中控制Eh(避免迁移)
盐度	G，(S)	含盐量	增加回灌淡水体积
养分	S，(G)	氮	对水进行预处理(如活性污泥法)
有机物	S，(G)	农药	排除有此可能的支流集水区
浑浊度和颗粒物	S，(G)	悬浮固体	对水进行预处理(如湿地)
放射性物质	G，A，(S)	α–射线	含水层的选择(避免)

注：表中A为含水层矿物；G为地下水；S为回灌的水源；来源一列带括号的表示可能的第二种来源；Eh为氧化还原电位(氧化还原倾向)。

对人类健康或环境的主要危害影响为：

- 衰减带以外的含水层(由此影响其他地下水用户和依赖地下水的生态系统)；
- 开采水的使用；
- MAR水处理与运行副产物的再利用与排放(见表4.2)。

表4.2　可管理的含水层补给运行导致有害副产物举例

一般危害	特殊危害举例
水处理副产物	任何产生废弃水(如反渗透)或副产品(如凝结作用、过滤、回洗水)的过程都可能会导致水体悬浮固体、病原体、无机物、养分、含盐量和有机物浓度的增加
净化水	补给水中悬浮固体、病原体、金属、养分和有机物可能在维护过程中在含水层储存和开采井中净化的水中富集
水池碎屑	补给水中的病原体、金属、养分和有机物可能在渗滤槽产生的碎屑中富集。如果符合质量标准，碎屑作为土壤改良剂的一种成分在农业中可以被重新利用

注：废物管理体系应优先采用以下次序：避免、减少、重复使用、再生、处理和去除。

发展一项 MAR 工程的倡导者应该请专家顾问来鉴别和量化健康、环境(如水文地质学和岩土技术)和管理风险。下一步就是采取预防性措施以减轻风险，优化监测方法以评估每次危害，包括在可适用的地方建立临界控制点。危害预防措施可能包括以下一个或多个屏障：

- 水源的选择；
- 补给控制系统(例如当监测指标值超出了临界控制限制，则停止补给)；
- 含水层的选择；
- 工程选址(远离敏感的依赖地下水的生态系统或最终用途)；
- 处理补给水以去除危害或其形成发生的前提；
- 需足够停留时间在含水层内进行被动处理；
- 在分配到最终用户之前对开采水的处理；
- 项目反馈计划，包括实时监测的反馈。

在许多 MAR 实施中，多重屏障是必需的，当其中任意一个屏障失败时，人类健康和环境依然能得到保护。表 4.3 总结了一些预防措施，其中许多措施应用于多种危险。依赖

表 4.3　MAR(危害)的预防措施和临界控制点的总结

<table>
<tr><th colspan="3">预防措施</th><th>描述</th><th>临界控制点</th></tr>
<tr><td rowspan="6">排除屏障</td><td rowspan="3">防止进入</td><td>危险源头控制</td><td>在补给之前水源的选择或管理(如集水处、开采水、屋顶径流、洪水)</td><td>否</td></tr>
<tr><td>摄入水平</td><td>在水下通过保持摄入水平排除漂浮污染物</td><td>否</td></tr>
<tr><td>排除没有达到临界限制的水</td><td>对指示变量的持续监测，当超过临界极限时提供水流转向或停止补给的反馈</td><td>可能性：依赖于危害、风险防范和监测系统</td></tr>
<tr><td rowspan="3">移除危害</td><td>土壤或含水层中的滞留时间</td><td>所有的人类病原体和所选的有机化学材料的减少</td><td>是：系统需要提供满足要求的滞留时间。开采速率受到限制以保证在补给和开采之间有足够的时间</td></tr>
<tr><td>在含水层中的运移距离</td><td>在实施情况的变动范围下，运移距离(在补给和开采之间)的选择提供一个最小的滞留时间</td><td>否：系统设计的一部分</td></tr>
<tr><td>处理过程</td><td>在补给之前或开采时，特殊危害的浓度降低</td><td>可能性：在补给点和(或)开采点；依赖于危害或对特殊环境终点的影响</td></tr>
<tr><td colspan="2" rowspan="5">在试运转和确认监测中控制危险的预防措施</td><td>预警反馈系统的实施</td><td>监测附近补给井的处理效率以进行预警，并允许实施纠正措施</td><td>是：探测超过临界限值以触发正确行动</td></tr>
<tr><td>补给水的开采和重新处理</td><td>开采污染水以防暴露</td><td>否：然而，通过预警系统减少开采水的体积</td></tr>
<tr><td>防止开采水用于无法接受的用途</td><td>如果开采水没有达到预期用途的水质标准，停止开采(允许更长的滞留时间)，转向可接受的用途或重新处理</td><td>可能性：与水质有关的代理参数可能用到</td></tr>
<tr><td>开采水的后处理</td><td>处理开采水以移除可鉴别的危害</td><td>是：直接或代理参数超过临界极限引发纠正措施</td></tr>
<tr><td>减小补给或开采的速率</td><td>在薄的弱含水层中限制流动速率以增加滞留时间，降低压力梯度，或减少 MAR 导致的速率和电平变动</td><td>否：服从于验证监测</td></tr>
</table>

于特定的工程，其他的临界控制点和预防措施可能更合适些。在执行和维持预防措施时，教育和培训是很重要的组成部分。

4.1　病原体

在承压含水层中，肠病原体的来源受限于它们在补给水中的数量。然而，在非承压含水层中可能存在其他来源。控制含水层补给的潜在水源(如废水、灰水、雨水)也包含许多威胁人类健康的病原体。

与控制含水层补给的和病原体有关的公众健康及环境风险与澳大利亚水开采指导方针(NRMMC-EPHC-AHMC, 2006)中的描述一致，如表 4.4 所示。

表 4.4　在未处理的污水中涉及到的微生物(NRMMC-EPHC-AHMC, 2006)

病原体类型	例子	疾病
细菌	沙门氏菌	肠胃炎，反应性关节炎
	弧状菌	肠胃炎，综合病症
	致病性大肠杆菌	肠胃炎，溶血–尿毒症综合征
	志贺氏杆菌	痢疾
	耶尔森氏鼠疫杆菌	肠胃炎，败血病
	霍乱弧菌	霍乱
	非典型的分支杆菌	呼吸道疾病(过敏性肺炎)
	军团菌	呼吸道疾病(肺炎，庞蒂亚克热)
	金黄色葡萄球菌	皮肤、眼、耳传染病，败血病
	绿脓杆菌	皮肤、眼、耳传染病
	幽门螺杆菌	胃溃疡
病毒	肠道病毒	肠胃炎，呼吸道疾病，神经错乱，心肌病
	腺病毒	肠胃炎，呼吸道疾病，眼传染病
	轮状病毒	肠胃炎
	诺如病毒	肠胃炎
	甲型肝炎	传染性肝炎
	杯状病毒	肠胃炎
	星状病毒	肠胃炎
	冠状病毒	肠胃炎
原生动物	隐孢子虫	肠胃炎
	贾第鞭毛虫	肠胃炎
	福氏内格勒	阿米巴脑膜炎
	痢疾内变形虫	阿米巴痢疾
寄生虫	绦虫	绦虫(牛麻疹)
	蛔虫	蛔虫
	鞭虫	鞭虫

4.1.1 MAR 中病原体的命运和归趋

地下水中病原体的生存受物理、化学和生物过程的影响(Toze 和 Hanna, 2002; Gordon 和 Toze, 2003)。含水层中病原体的失活凸显了它们在多屏障方法中作为耐用处理屏障的作用。病原体在含水层中的存在和生存是非常易变的，受多种因素的影响，包括：

- 病原体的类型；
- 补给水源类型；
- 温度；
- 氧化还原态和氧气浓度；
- 当地地下水微生物的活性；
- 含水层地球化学性质。

含水层中这些参数影响病原体生存的更多信息已由 Dillon、Toze(2005)和 NRC(2008)给出。尽管所有的参数(独立地或全体地)可能影响病原体的生存，但是不同病原体的环境稳定性(如在不同的地区)也不同，这依赖于当地地下水条件。肠道病毒通常被认为具有最强的抗衰减能力(例如具有最高的生存潜力)；接下来是原生动物，然后是细菌。

人们对 MAR 期间寄生虫的归趋知之甚少。在澳大利亚(除北部热带地区)，寄生虫具有很小的危险性。它们可以通过适当的预处理得到控制，并且在稳定的池塘中通过简单的处理系统，如凝结、絮凝和沉淀，可以有效地从水中去除(Jimenez，2003)。即使一个寄生虫卵能够通过处理系统，它们的大小(40~90 μm)也使得其在更加压实的砂质含水层中通过简单的过滤过程就可以把它们去除。

可管理的含水层补给工程有赖于含水层处理以去除病原体，在开采饮用水之前，供给水总是需要得到这方面的确认。病原体的去除效率依赖于当地的特殊情况，通过监测确认衰减速率可以分辨不确定因素。根据对数去除率，含水层在概念上更像其他天然和工程的病原体水处理过程(NRMMC-EPHC-AHMC，2006)。含水层的对数去除率主要与补给水的滞留时间、含水层的氧化还原态和温度有关；通常，根据病原体数量减少一个对数所需的天数来表述。表 4.5 说明了减少一个对数的时间范围内，选定的病原体及在需氧和厌氧含水层中发现的指示生物。

表 4.5　需氧和厌氧含水层中微生物失活速率举例

含水层氧化还原态①	减少一个对数的时间(d)					
	腺病毒	轮状病毒	贾第鞭毛虫	似隐孢菌素	大肠(杆)菌	弧状菌
需氧	35~87	112②	—	22②	1~3	—
厌氧	53~70	72~118	—	3-89	4~11	—

注：①这里作为例子给出的结果是从 MAR 计划里得到的，它的地下水温度高于 20 ℃。任何温度低于 20 ℃的依赖于病原体灭活的含水层，需要有病原体生存评估；

②单个研究结果；

③来源：修改来自 Toze 等(2002)、Gordon 和 Toze(2003)、Wall 等(2004)、Sidhu 等的研究。

在表 4.5 所示的病原体的失活速率相对较广，因为它们受上述地点特殊情况影响。更多特殊病原体的生存数据在附录 1 中给出。

按照经验，在给定地点对病原体衰减最可靠的评价是测量原地接种扩散室病原体的存活量。由于水源中病原体数量的时间变化，在地下水样品中没有检测出病原体，不能阐明病原体的实际衰减速率。有关实验室得到的衰减速率必须是在与含水层有关的温度和氧化还原态的情况下，用来自含水层的微生物的接种室得到的。

普通的数据，如表 4.5 给出的，适用于剩余危险预评估。随着来自 MAR 场所有效数据的积累，估算含水层病原体衰减速率有望得到更高的精度。

检验监测包括检测开采水中的微生物指示和靶向肠道病毒。这是通过每月在衰减地带中观测井和开采水取样做的。

在含水层中二次处理污水似乎不影响本土病原体的数量。实验室评价显示：二次处理污水对本土含水层物质中病原体含量的影响与控制相比不会有大的区别(S.Toze，CSIRO Land 和 Water, pers comm., 2007)。

最近研究(Reed, 2007; Reed 等, 2007)记载了玻利瓦尔附近再生水 ASR 实验和 Floreat Park 再生水渗水廊道中天然微生物种群的变化。这些研究证实了与接近富营养化的补给水相关的微生物数量的变化趋势和氧化态的变化。在那些水质能表明存在开采水的末梢水中，微生物群落的数量、生物多样性和活跃性是不变化的。尽管证据不具有决定性，然而它仍然表明 MAR 不会比通过抽取天然地下水使本土的致病菌(如绿浓杆菌，气单胞菌属)机会恢复到更高的数量。

4.1.2　MAR 对病原体的管理

降低病原体危害风险和达到执行目标的预防措施包括：

- 水源控制(如集水流域雨水水源管理)；
- 通过处理过程去除病原体(如用工程或天然处理过程获得所需的参照病原体的对数去除率)；
- 通过就地预防措施减少暴露(如在用再生水灌溉的过程中控制公共入口)。

上述三种措施在澳大利亚水再循环指导方针的第 3.4.1~3.4.3 部分以及附录 3 中都有描述(NRMMC-EPHC-AHMC, 2006)。

可容忍风险的概念是通过管理含水层补给对肠道病原体进行管理的核心。这些指导方针采用了每人每年 10^{-6} 伤残调整期望寿命(DALYs)的可容忍危险(NRMMC-EPHC-AHMC，2006)。

基于潜在大量病原体危害的假定，选择三种参照病原体——轮状病毒、隐孢子虫属和弯曲菌分别代表病毒、原生动物和细菌。关于 DALYs 的详细描述和基于微生物健康的参照病原体执行目标的计算可参照澳大利亚水再循环指导方针的附录 2(NRMMC-EPHC-AHMC，2006)。MAR 唯一特性就是考虑了含水层对病原体的影响。

4.2　无机物

本部分适用于主要离子 (钙，镁，钠，钾，氯化物，硫酸盐，重碳酸盐，溴化物和氟化物)；金属 (铝，镉，铬，铜，铁，锰，镍，铅，锶和锌)；类金属(砷，硼，硅)；气体(硫化氢和甲烷)。氮和磷将在第 4.4 部分单独讨论。

和病原体不同，没有足够的关于化学参数的信息支持 DALYs。因此，可容忍风险是按照指导浓度定义的(ANZECC 和 ARMCANZ，2000；NHMRC-NRMMC，2004；WHO, 2006；EPHC-NHMRC-NRMMC, 2008a).

含水层储存导致的主要无机危害有：

- 砷、铁、锰、微量核素或硫化氢增加，产生的开采浓度超过有益利用限定值；
- 开采水中铁离子含量增加，对给水工程基本设施产生影响(如灌溉)；
- 主离子化学的变化改变开采水中盐或养分平衡，从而影响预期使用的适宜性(如灌溉)。

当来自注入井的反洗水，或含水层储存开采井的原生水的金属含量的增加超过了宜用性限定值，应该考虑处理、利用或处置这种废液。

4.2.1 MAR 中无机物的来源与归趋

MAR 含水层水的化学性质受化学反应影响，受含水层条件(如 pH 值、氧化还原态、矿物质、有机物质、微生物活性)的驱动。源水与天然地下水、源水与含水层介质间都可发生反应。水质和含水层的渗透性由此产生变化。与地下反应有关的关键性危害描述如下。

4.2.1.1 砷的增加

如果蓄水区中的黄铁矿被氧化，或者溶解了 3 价铁的氧化物，含水层沉积物中砷将溶出(见附录 3 的决策树)。这对于目前矿物正在减少的限制目标区十分关键，尽管初始时刻源水的砷浓度在可接受范围内(亚瑟等，2003 年)，也可能导致砷浓度大于饮用水指标。

4.2.1.2 铁的增加

蓄水区沉积物中铁的释放主要发生在源水中的有机物与 3 价铁的氢氧化物或氧化物(如针铁矿，赤铁矿)发生反应时。黄铁矿氧化以及蓄水区 pH 值的改变也会释放出铁(见附录 3 的决策树)。铁的释放通常影响水的感官，导致水的颜色加深；也会导致含水层堵塞。铁的增加也与其他危险物质如砷和镭的释放有关。

4.2.1.3 锰的增加

当与源水中的有机物反应，或者蓄水区的 pH 值改变时，沉积物中的氧化锰和氢氧化锰会释出锰(伊比森等，1994)。与铁一样，锰也会使开采水产生颜色。

4.2.1.4 微量离子的增加

受影响的离子种类有：铝、锑、砷、钡、铍、镉、铬、钴、铜、氟、铁、铅、锰、钼、镍、钒、铀和锌。所有这些离子都会带来健康与环境问题。微量组分的增加通常与铁、锰和砷的增加相一致。

微量离子释放的机制包括：

- 硫化矿物的氧化，比如黄铁矿，其中存在微量元素(与砷类似，尽管砷一般比阳离子具有更强的迁移性)；
- 3 价铁的氧化物因 pH 值或 Eh 值的改变而溶解，其表面通常包含被吸附的微量核素；
- 在固体表面与其他离子发生交换或置换(如阳离子交换)；
- 矿物平衡，当蓄水区中的水与优势矿物不平衡时；
- pH 值或 Eh 值的改变引起矿物相的溶解或微粒积累。

4.2.1.5　硫化氢的增加

有机物进入缺氧的蓄水区与溶解硫酸盐反应产生硫化氢气体。在以微生物为媒介的一系列氧化还原反应中，铁的反应可能先于硫化氢的生成。硫化氢通过味道和气味对开采水产生感官危害。

4.2.1.6　主要离子成分变化

源水与地下咸水混合，或与固相物交换位发生离子交换，都能明显地影响钠、钙和镁的含量。如果恢复后的水用于灌溉，则这一因素不容忽视，因为它能改变土壤碱度的危害程度。

4.2.1.7　碳酸盐矿物的过度溶解

碳酸盐矿物的过度溶解会：

- 影响注水井和弱透水层的稳定性；
- 增加开采水中的含沙量；
- 延长优先流路径，减少了可以降低危害的停留时间。

碳酸盐的溶解同时也可能揭露如硫化物的反应面，并且增加金属和非金属释放的可能性。

MAR 方案中使用的源水不太可能与蓄水层中的矿物达到平衡。结果，当源水接触到含水层中的矿物时将使其发生溶解。溶解程度决定于矿物在给定条件(如 pH 值、温度、压力、离子强度、接触时间)下的溶解度。在地下蓄存期间的矿物平衡也可因其他反应过程而发生变化。例如，在细菌还原硫酸盐导致溶解硫酸盐浓度降低后，含水层沉积物中重晶石($BaSO_4$)的溶解释放出钡(Zhou 和 Li，1992)。

碳酸盐矿物是影响开采水水质的主要因素，因为碳酸盐溶解是一个快速反应。相比之下，pH 值在 6~8 时硅酸盐的溶解是一个非常缓慢的过程(Appelo 和 Postma，1999)，并且在可管理的人工回灌方案的时间尺度上对水质的影响微乎其微。碳酸盐矿物的溶解可以增加含水层的渗透性，而且对注水井和弱透水层稳定性的影响不可忽略。

矿物溶解增加了可开采水的盐度和硬度，并且增加了诸如钡和氟(萤石的溶解)等次要组分。溶解度控制可以限制一些有害物质的浓度，如钡(重晶石溶解度)，磷酸盐(磷灰石溶解度)或者铁(铁的氢氧化物或氧化物)等，但是矿物的沉淀能带来阻塞问题。矿物溶解或析出的趋势可以通过饱和指数检验。

蓄水层中的氧化还原态改变了开采水中的无机化学。MAR 中的氧化还原反应通常在含氧的源水进入缺氧的含水层，或有机物进入含水层时发生。在进行可管理的含水层补给操作的过程中，氧化还原带及相应的水质具有时空变化。高度反应区通常在注水点或渗透点产生，导致与大量储存水水质的差异。例如，如果源水中有机物含量较高，反应就变成了厌氧的。这可以导致铁的溶解(或者出现在含水层沉积物中，或者源水中过滤出的颗粒物质在注水井附近的积累)或原位硫化氢的产生。

氧化还原条件和水质的变化也可以在下列情况下发生：

- 随时间推移，流量不同(如相对蓄水量的注水量或入渗量)的情况下；
- 饱和度改变(如土壤含水层处理的潮湿与干燥周期)；
- 含水层储存与回采操作过程中的回流。

黏土矿物、有机物或铁氧化物的表面吸附作用可以减少痕量金属和非金属。然而，吸附不是永久的。其他物质的优先吸附，或者由 pH 值、Eh 值决定的表面性质的变化所引起的可吸附点数量的变化都会导致解吸。另外，含水层的吸附容量也会被限制，使得对有害物质向下游侵入的监测或恢复定位被延误。

如果天然地下水为半咸水或有害物质的含量超过了特定资源化利用的目标值，则混合对开采水的水质具有重大影响。两种水混合可以产生对含水层矿物质侵蚀性更强的溶液(Runnels，1969)。如果地下水中的一种无机组分超过了水质标准，当别的过程，比如矿物质溶解、释放来自含水层沉积物的物质时，简单的稀释将不足以保证满足水质标准。在盐度目标和操作规范中应考虑混合效应。

源水中的金属大部分以微粒态存在，并在地下靠近入水处(如井口或盆底)积累。积累的金属可以通过运行维护，如井的改造或盆地废料场，被永久地从系统去除。一些原位溶解也可导致局部金属含量增加，这很可能影响通过蓄水层和开采井恢复的水。

4.2.2 无机化学管理

源头控制应包括限制有害行为(如集水管理，废水排放许可)和水质标准之外的流动(如 pH 值，传导率)。

预处理方法包括：

- 源水处理(如过滤，凝结，絮结)；
- pH 值调节(如防止锰从沉积物中析出，伊比森等，1994)；
- 控制氧化还原条件限制含水层中的反应(如限制并氧化原水中的有机碳以防止砷释放)。

含水层无机化学材料的残留风险评估要建立在正确理解化学过程的基础上，并且明确管理人类健康和环境风险的必须预防措施。一些矿物尽管只是微量存在，但仍具有危害，所以必须要对其进行监测。在确定有回灌水进入的含水层中所含物质的时候，岩心与第 2 阶段中采集的地下水水样是不能使用的。

以复原为目的或者在必要的地方为达到更高的使用标准引起咸水化的主要离子应该被移除。这可以通过后复原处理实现，比如通风和过滤，或者其他技术(这些方法可以单独使用或配合使用)，包括氧化、沉淀、凝结、吸附、离子交换、石灰软化和过滤(NHMRC-NRMMC，2004)。

4.3 盐度与钠质化

与 MAR 相关的与盐度和钠质化(相对钙、镁的钠的含量)有联系的公共健康和环境风险包括：

- 盐度值超出总溶解盐或钠含量的资源化利用值；
- 咸水灌溉产生的渗透作用对作物生长和产出的影响；
- 淋溶消除过量盐度的用水导致水位上升；
- 与钠质化有关的耕地土壤结构衰退；
- 盐度对基础设施和其他设备产生影响(如管道、配件、仪表锈蚀，砌石盐类吸潮)。

在 MAR 中，补给水与周围地下水混合将导致开采水盐度不同于补给水。一般情况下，作为含水层内可管理含水层补给对象的地下水的含盐度应该接近或略高于源水(与地下水保护方针中的原则一致，ANZECC-ARMCANZ，1995)。因此，天然地下水成为开采水盐分和钠的另一来源。澳大利亚水循环指导方针中讨论了盐度和钠质化的环境风险以及它们对土壤结构和农业生产的影响(NRMMC-EPHC-AHMC，2006)。

4.3.1 MAR 中盐分的来源

MAR 所有种类的源水都存在天然盐分水平，这些盐多来自于无机盐，少量来自于有机物溶解和小的胶体物质。源水中的无机物成分可以通过电导、总溶解盐或钠质化等计量方式来描述。通常，屋面径流中的盐度低于暴雨水径流(的盐度)。暴雨水径流的盐度又低于来自污水的循环水。这是因为水在流经农村和城市地区时有自然或人工过程富集或加入了盐分，导致含盐程度增加。

地下水的含盐程度从淡水到高含盐量不等。浅层地下咸水渗透进入污水管道使污水的含盐度和含钠度持续增加，除非经过处理或混合，这些水不适合用于可管理的含水层补给的再循环。

4.3.2 盐度与钠质化管理

管理控制包括下列预防措施：

- 排水区水质管理和水源控制(最小化盐分输出，消除或在可行的情况下减少盐度的点源)；
- 源水选择；
- 选择风险最小的目标含水层；
- 预处理或后处理措施(脱盐作用)；
- 将开采水与其他低含盐量水混合。

4.4 养分：氮、磷和有机碳

本部分讨论氮、磷(水循环的环境有害物质，NRMMC-EPHC-AHMC，2006)和有机碳(与地下微生物作用过程有关的一种重要的养分)。

由于可能引起灌溉水的养分不平衡、土壤富营养化和导致陆地生物的毒性效应，氮和磷被认为是关键的环境危害物质(NRMMC-EPHC-AHMC，2006)。尽管地下蓄水可能降低养分的含量，但是开采水的整体营养平衡仍需结合资源化利用来考虑。源水中的养分(主要是有机物)会刺激地下微生物的活动。反过来，这也改变了水中无机物和有机物的含量并影响含水层的渗透性。

4.4.1 MAR 中养分的来源与归趋

源水中养分的水平与变化很大程度上受预处理方法的影响。开采水可能含有高浓度的养分，并随季节对微生物处理过程的影响而改变。暴雨水径流的养分含量通常比循环水低，但会随着汇水区的类型而改变(如工业区)。

在 MAR 提供的操作中，移除有机碳和氮是一种消极的水质处理方式(Dillon 和 Toze，2005)。有机物可以通过生物降解、微生物同化、过滤、吸附或沉淀等作用消除。生物降解通过氧化还原过程起作用，影响无机物的迁移与有机物的归趋。胶体有机物可以促进其他化学有害物的转移。通过生物降解消除的有机物的量取决于该有机物的性质；易降解的部分可以在进入地下后几天内消除，而反应性低的物质可能要经过一段更长的时间才能降解(Fox 等，2001)。沉淀物中的活性有机质也有可能降解。当富集养分的源水被引入地下时可发生微生物同化，并在切入点附近形成生物膜(如井口或盆底)。这层生物膜形成了一个与其他蓄水带不同的、具有特殊氧化还原过程的化学反应带，并对地下水水质处理起重要影响。颗粒有机碳的过滤为微生物的活动提供了额外的能量来源，并能产生水溶性降解副产品。

源水中的氮能以不同形态存在，包括有机氮(主要是蛋白质)、氨、硝酸盐、亚硝酸盐和气态氮。循环水中氮的主要形态是有机氮、氨和硝酸盐。氮的归趋决定于其形态和所处的氧化还原条件。在好氧条件下，硝化作用可将氨转变为硝酸盐；在厌氧条件下，氨可以通过离子交换作用被吸附在矿物表面，直至超过交换容量。硝酸盐可以被还原为氮气或氨而除去。在非饱和带，部分氨转变为氨气并在空气–水界面挥发而散失。

4.4.2 养分的管理

源头控制包括限制有害行为，以及在水质指标(如颜色和浊度)超过预处理或排放标准时改流。

预处理方法有：

- 源水输送设施的连线过滤(针对有机碳)；
- 生物过滤；
- 湿地被动处理。

天然处理系统(如湿地去除养分)的效率取决于它们的维护。监测对辅助管理湿地处理系统是必须的。

地下有机碳和氮的去除有赖于通过氧化还原过程进行处理(如土壤含水层处理；Fox 等，2001；Amy 和 Drewes，2007)。其有效性需要由浓度降低和生物化学条件的证据来证明。磷的去除也需要浓度降低的证明，可以由矿物学(铁、铝的氧化物)或矿物饱和度计算获得支持。如果需要高价值的使用，开采水中的有机碳可以通过粒状活性碳和薄膜渗透作用得到恢复(NHMRC-NRMMC，2004)。人工回灌井、渗透池和渗水廊道的生物堵塞也应加以考虑。

4.5 有机物

本部分讨论痕量有机化合物(通常指微量污染物)，包括：

- 农药；
- 碳氢化合物；
- 多环芳烃(PAHs)；
- 新兴的相关化学品；
- 内分泌干扰物；
- 个人护理用品；
- 药品；
- 阻燃剂。

痕量有机物主要是人为来源(比如多环芳烃就是碳的燃烧产物)；然而，也有一些是自然的产物(如藻毒素)。

有机化学品能够引起健康风险(NHMRC-NRMMC，2004)和环境风险(NRMMC-EPHC-AHMC，2006)。有大量新兴化学品，如内分泌干扰物、药品和个人护理用品，还有一些消毒副产品(如 N 亚硝基二甲胺，NDMA，NRMMC-EPHC-AHMC，2006)。如果对某种特殊化学品没有建立水质指标值，EPHC-NHMRC-NRMMC(2008a)建议所有化学品的含量均要符合饮用水的使用要求；同时它也提供了处理混合化学品的指导方针。

环境毒性检测可以提供更多的信息，帮助人们了解 MAR 项目对敏感环境所造成的影响。环境毒性测试指导在澳大利亚与新西兰水生生态系统淡水和海水水质指导方针的第三章(ANZECC 和 ARMCANZ，2000a)。

4.5.1　MAR 中有机物的来源和归趋

由于有机物间或地进入水体和分析检测能力的局限性，确定源水中有机药品的存在及进行相关的风险评估是很困难的。不过，可以根据源水的起源和有机药品存在的性质估计其可能性。确定源水中有机组分的特性必须与公众健康和环境风险挂钩。

通常情况下，地下蓄水提供了一种有机药品的处理步骤(Dillon 和 Toze，2005)。但是,已有了储水期三卤甲烷(一组消毒副产物)形成和衰减的报告(Pavelic 等,2005,2006b)。通过减少源水中有机物的含量，或改变消毒方式以减少余氯，消毒副产品产生的可能性可以降低(Jimenez，2003)。

在地下，有机化学物可以通过挥发(非饱和带)和生物降解(非饱和带与饱和带)去除。降解速率随 pH 值、温度、氧化还原态、共基质(如溶解有机碳)的存在而改变。吸附作用也会延缓有机物的迁移，这种吸附的去除能力有限，但却能够增加进行降解的时间。只吸附在氧化还原条件和与含水层蓄水带有关的温度下不发生降解的化学药品时的含水层衰减不应考虑。

有机化学品环境归趋数据的有无变化很大。关于有害物质如苯的归趋已有很多报道(Howard，1991)，但是关于新兴化学品的信息就很少。一个关键问题是保证现存的环境归趋数据是在考虑 MAR 方案的前提下，在相似的生物化学环境下测得的。在缺乏野外数据的情况下,实验室研究可以预测在设想条件下有害物质的归趋(Oliver 等, 1996; Ying 等，2003)。

如果蓄水区是一个非承压含水层，有机化学物可能来自点源(如工业行为)和面源(如杀虫剂的使用)。源水与天然地下水的混合可能会影响开采水的水质。

4.5.2　有机物的管理

源控制包括限制危险性活动、工业废水和暴雨水回灌，并改进高风险行为的危害管理，达到降低来源浓度、防止突发泄漏的目的。

预处理和后处理措施包括生物过滤、湿地被动处理和深度三级处理(如膜过滤)。

含水层有机物的衰减依赖于浓度的有效降低，而浓度的降低随着时间和距离增加，并依赖于含水层滞留时间及其物理化学和氧化还原条件。

选择不同的指示剂进行检测，关注能够反映有效系统性能最灵敏的指示剂(Drewes 等，2008)。

4.6 浊度和颗粒物

可管理的含水层补给中与浊度有关的公众健康和环境的风险包括：

- 具有浊度的开采水：

——超出饮用水标准(最终以饮用为目的)；

——若不去除则影响泵和灌溉基础设施。

- 减弱消毒效果，导致病原体微生物危害增加。
- 增加了多种可吸附的污染物传输的风险，包括：

——重金属；

——磷；

——多种有机物；

——病原微生物。

- 在注入井改造(反冲洗)期间，反冲洗后水的排放会影响下游暴雨水的收集。

4.6.1 MAR 中浊度的来源

用于可管理的含水层补给的所有源水均包含自然水平的颗粒物(以浊度或悬浮固体度量)，它们来自无机淤泥，黏土级粒度的颗粒物和有机物。暴雨径流的浊度通常变化很大，这与气候、流域地貌和土地的使用管理等因素有关。典型的经过二次或三次处理的废液与暴雨水径流相比，含有较低的颗粒物浓度和较高的有机物含量。典型的屋面径流颗粒物含量低，但是由于植物碎屑沉积或管理不善也会含有较多的颗粒物。除非水井的设计改造不当，地下水的浊度一般不高。

土壤或含水层中的矿物分解和颗粒物的运移，以及通过反冲洗注水井保持补给速度的常规操作，都会导致可管理的含水层补给在实际中产生颗粒危害。

4.6.2 浊度管理

浊度管理控制包括如下预防措施：

- 水源选择。
- 流域水质管理和来源控制：

——最小化微粒输出；

——消除或削弱浊度的点源(可行的)。

- 回灌前通过下列措施进行预处理或后处理：

——沉降槽；

——湿地；

——凝结；

——过滤。

4.7 放射性核素

环境中存在天然的放射性物质(如铀、钍、钾 40)，人类暴露于辐射的危险主要来自自然界。另外，人类也可能因医药(放射性药品)和工业使用放射性物质等人类活动而受到辐射。

对放射性核素的关注主要是因为开采水会导致人体健康风险，因为人们通过粮食作物、畜牧用水或食物链积累摄取饮用水或食物(镭和氡)，或者吸入释放自供水的气体(氡)。

放射性以贝可(Bq)为单位，1 Bq 为 1 s 内发生的一次核裂变。人体健康的考虑因素是根据有效放射剂量，以西韦特(Sv)为单位，认为所有组织器官接受相同剂量的辐射，并考虑它们对放射性敏感度的不同进行加权。通过饮用水所能接受的放射性剂量应<1 mSv/a(NHMRC-NRMMC，2004)。摄入每单位放射性核素的放射性剂量可以使用澳大利亚饮用水指导方针中表 7.1 和第 7.6 部分估算(NHMRC-NRMMC，2004)。

4.7.1　MAR 中的放射源

接受了来自医药和工业用水的循环水或暴雨水可能会含有放射性核素。地下水可能含有天然形成的镭和氡的同位素(镭 226、镭 228、氡 222)。采矿活动也可能富集天然形成的放射性核素(如处理矿砂，生产磷酸盐废料)。

MAR 中放射性核素的主要来源是含水层蓄水过程中蓄存水与含水层基质间的相互作用。天然地下水中的放射性元素可以用来指示开采水放射性的最低水平。天然的放射性核素浓度差别很大，取决于下列含水层的性质(Dillon 和 Toze，2005)：

- 地质状况；
- 孔隙率；
- 颗粒大小；
- 氧化还原态；
- 主离子化学。

一般情况下，在断裂花岗岩(结晶的)含水层和靠近富含有机煤沉积的地方，放射性核素浓度高(Herczeg 和 Dighton，1998)。也有人报道了来自碳酸盐含水层的铀淋滤(Williams 等，2002)。

在进行可管理的含水层补给之前，开采水和天然地下水中的氡含量接近，这是因为含水层介质与源水间的氡在不到三周的时间内就可以达到平衡(Dillon 和 Toze，2005)。

铁和锰的氢氧化物可以吸附大量的镭。因此，在 MAR 方案中如果有机物的氧化导致铁或锰的氢氧化物表面溶解，镭的浓度就会升高。镭的浓度也能通过含镭矿物如磷酸盐的溶解而升高(Dillon 和 Toze，2005)。

4.7.2　放射性核素的管理

如果总的 α 粒子和 β 粒子超过了目标值，蓄水带天然地下水的放射性可以通过测量总的 α 放射性和 β 放射性(不包括钾 40)，进而分析单一的放射性核素而被检测出来(NHMRC-NRMMC，2004)。

如果目标含水层是镭的潜在来源，地球化学模型可以通过矿物溶解和有机沉积物氧化评估额外释放的能力。模型同样需要定义源水中合适的 pH 值、氧、硝酸盐和有机碳含量，以最小化含水层蓄水过程中地球化学反应释放镭的能力。可能使用的预处理和后处理措施有氡 222 通风、石灰软化、离子交换，或镭 226 和镭 228 的反渗透(NHMRC-NRMMC，2004)。

附 录

附录 1 可用于调查与核实检测的方法实例，组成第 3 阶段的调试与实验

附表 1 提供了可用于指导第 3 阶段评估的工具实例，包括核实检测在内。针对不同的有害物给出实例。概括描述每种方法的信息来源与案例研究，并指明参考文献。

附表 1 可用于指导第三阶段评估的工具实例

危害	分析方法	提供资料	参考文献
病原体	文献	指示特定病原体或指示剂的源浓度和衰变率	Appendix 2 and NRMMC-EPHC-AHMC (2006)
	现场或实验室衰变研究	特定点的衰变率	Pavelic 等(1998)
	原水与天然地下水取样分析[①]	背景地下水源浓度和理化性质	Toze 和 Hanna (2002) Toze 等 (2002)
	示踪试验，含水层泵试验，电磁流量计[①]	地下水流速和流动时间	Pavelic 等 (2006a)
无机物	文献	源浓度指示	Appelo and Postma (1999)
	决策树	过程定性	Appendix 3
	原水与天然地下水取样分析[①]	背景地下水源浓度和理化性质	Herczeg 等 (2004)
	土壤与含水层取样分析	存在度和数量活性周期	Bekele 等(2007)
	实验室级分栏和分批测试	动力学过程	Bekele 等(2007)
	地球化学模型(PHREEQC[②]，PHT3D[③])	过程定量理解	Greskowiak 等(2005)
盐度和钠度	文献	源浓度指示	NRMMC-EPHC-AHMC (2006)
	原水与天然地下水取样分析[②]	背景地下水源浓度和理化性质	Herczeg 等(2004)
	土壤与含水层取样分析	活性黏土和盐负荷	Bekele 等(2007)
	实验室级分栏和分批测试	操作设计协助	Bekele 等(2007), Cavé (2000)
	地球化学模型 溶质运移模型	地下混合级别	Pavelic 等(2006a)
营养质	文献	源浓度指示	NRMMC-EPHC-AHMC (2006)
	原水与天然地下水取样分析[①]	背景地下水源浓度和理化性质	Vanderzalm 等(2006)
	土壤与含水层取样分析	存在度和数量活性周期	Vanderzalm 等(2006)
	实验室级分栏和分批测试	操作设计的运转，传输和协作	Fox (2002)

续附表 1

危害	分析方法	提供资料	参考文献
有机化学	文献	指示源浓度与环境变化数据	NRMMC-EPHC-AHMC (2006)
	原水与天然地下水取样分析	周围地下水源浓度和理化性质	Pavelic 等(2005)
	土壤与含水层取样分析	存在度和数量活性周期	Pavelic 等(2005)
	被动采样器	综合源浓度低水平探测	Komarova 等(2006)
	实验室级分栏和分批测试	操作设计的运转，传输和协作	Ying 等(2003)
	地球化学模型(PHREEQC②, PHT3D③)	过程定量理解	Greskowiak 等(2006)，Miller 等(2002)
	示踪试验，含水层泵试验，电磁流量计	地下水流速,运移时间	Freeze and Cherry (1979)，Clark 等(2005)
浊度和颗粒物	文献	指示源浓度	NRMMC-EPHC-AHMC (2006)
	原水与天然地下水取样分析①	背景地下水源浓度和理化性质	Pavelic 等(2006b)
	土壤与含水层取样分析	当前的反应量	Rinck-Pfeiffer 等(2000)
	实验室级分栏和分批测试	操作设计的运转，传输和协作	Rinck-Pfeiffer 等(2000)
放射性核素	文献	浓度指示	Cook and Herczeg (2000)
	原水与天然地下水取样分析①	背景地下水源浓度和理化性质	Popit 等 (2004)
	土壤与含水层取样分析	沉积物反应	Gundersen and Wanty (1993) Herczeg and Dighton (1998)
水力：压力，地下水位，扩散，弱透水层稳定性	文献	初始参数估计	Freeze and Cherry (1979)
	地下水水位监测	静水条件	Freeze and Cherry (1979)
	含水层抽水试验①	局部含水层水力特性	Freeze and Cherry (1979)
	物理地球测井	含水层禁闭，完整性	Pavelic 等(2001)
	地下水模型	量压变化情景模型	Dillon 等(1999)
	土工技术评价	弱透水层岩土性质	Jones 等(2005)
	地球化学评价和建模	矿物质溶解对水井稳定性的影响	Le Gal La Salle 等(2005)
依赖地下水的生态系统	文献	初始参数估计	Freeze and Cherry (1979)
	地下水水位监测	静水条件	Freeze and Cherry (1979)
	原水、天然地下水和地表水的取样分析①	背景地下水源浓度和理化性质	Fies 等(2002)
	液压和溶质运移模型	量压波动及其对地表水影响的情景模型	Fies 等(2002), Escalante 等(2005)
	生态毒性测试	环境影响	ANZECC-ARMCANZ (2000a)

注：EM=电磁。

①可能需要钻探新井；

②设计计算机程序进行低温水地球化学计算；

③在饱和多孔介质中活性多组分运动模型。

附录 2　含水层中病原体的失活速率

附表 2 表明了原位扩散室的研究和实验室研究中，地下水中不同的肠道微生物的衰减率。这些速率用来计算不同的 MAR 类型和装置中病原体的衰减。计算的结果可用来指明地下水中不同的肠道微生物存活的持久性，而且这些只能用来作为一个指导。所有的结果都是从存在活跃的地下水微生物的实验中获得的。

附表 3 给出了被用于实验研究的不同肠道微生物的简单描述。如果病原体的残余风险不能令人接受，那么就需要考虑采用消毒措施(在有沙门氏菌风险的条件下)，通过利用现场悬浮在井中的多个容器，能够确定出修正后的衰减率的大小。

附表 2　肠道微生物的衰减时间(T90)(取对数)

微生物	温度(°C)	好氧型		厌氧型	
		T90(d)	参考资料	T90(d)	参考资料
柯萨奇病毒[a]	15	10	Gordon 和 Toze，2003[b]	—	—
柯萨奇病毒[c]	19~21	20	Jansons 等，1989[d]	—	—
柯萨奇病毒[a]	27	8	Toze 和 Hanna，2002[d]	—	—
柯萨奇病毒[a]	28	10	Gordon 和 Toze，2003[b]	—	—
柯萨奇病毒[a]	28	7~9	Toze 等，2002[d]	144	Gordon 和 Toze，2003[b]
变形链球菌 2[a]	12	77	Collins 等，2006[be]	—	—
变形链球菌 2[a]	12	6~33	Yates 等，1985[bf]	—	—
变形链球菌 2[a]	13	9~13	Yates 等，1985[bf]	—	—
变形链球菌 2[a]	15	1	Gordon 和 Toze，2003[b]	—	—
变形链球菌 2[a]	17	13	Yates 等，1985[bf]	—	—
变形链球菌 2[a]	18	12	Yates 等，1985[bf]	—	—
变形链球菌 2[c]	22	6	Toze 等，2002[d]	—	—
变形链球菌 2[a]	23	<1	Blanc 和 Nasser，1996[be]	—	—
变形链球菌 2[a]	23	2~6	Yahya 等，1996[bf]	—	—
变形链球菌 2[a]	23	2~5	Yahya 等，1996[bf]	—	—
变形链球菌 2[a]	27	5	Toze 和 Hanna，2002[d]	—	—
变形链球菌 2[a]	27	<1	Alvarez 等，2000[bf]	—	—
变形链球菌 2[a]	28	3	Gordon 和 Toze，2003[b]	8	Gordon 和 Toze，2003[b]
变形链球菌 2[a]	28	8~9	Toze 等，2002[d]	—	—
脊髓灰质炎病毒[a]	15	5	Gordon 和 Toze，2003[b]	—	—
脊髓灰质炎病毒[a]	28	1	Gordon 和 Toze，2003[b]	77	Gordon 和 Toze，2003[b]
大肠杆菌[a]	15	1	Gordon 和 Toze，2003[b]	—	—
大肠杆菌[c]	22	3	Toze 等，2002[d]	—	—
大肠杆菌[a]	27	1	Toze 和 Hanna，2002[d]	—	—
大肠杆菌[a]	28	1	Gordon 和 Toze，2003[b]	6	Gordon 和 Toze，2003[b]
鼠伤寒沙门氏菌[c]	22	7	Toze et al 等，2002[d]	—	—
鼠伤寒沙门氏菌[a]	27	1	Toze 和 Hanna，2002[d]	—	—

注：a 表示实验研究所确定的存活比率；
b 表示地下水中病菌衰减的研究，并不一定是关于特殊可管理的含水层补给方案；
c 表示现场研究确定的存活比率；
d 表示特殊的可管理的含水层补给方案中的一部分研究；
e 表示期刊中并没有指明氧化还原条件，也没有提供确定后来氧化还原条件方法的足够的信息；
f 表示期刊中并没有指明氧化还原条件，但从所用的实验方法中可使人猜想到氧化条件。

附表 3 肠道微生物的简单描述

肠道微生物	详细资料
柯萨奇病毒	一种与脊髓灰质炎病毒密切相关的肠道病毒， 而且是人类群落中，尤其是年轻人群中的一种普遍的病菌 能引起手、足、口方面的疾病 已经证实，在大多数条件下，比脊髓灰质炎病毒有更高的抵抗力 一种专性细胞间病菌
大肠杆菌	一种普遍用于指示排泄物和治疗效果的肠道微生物 一种能够产生毒素的病菌 在水中代谢旺盛
变形链球菌 2	一种只能影响特定种类的大肠杆菌细胞的病毒 对人类或动物不致病 经常作为肠道病毒的替代物,原因是其相似的形态以及实验和研究的易获取性
脊髓灰质炎病毒	能产生小儿麻痹症的一种肠道病毒 由于在社区中进行了广泛的疫苗接种，在澳大利亚并不会引起严重的健康危害，但是疫苗经常在下水管道中被发现 一种对环境的抵抗力最小的一种肠道病毒 一种专性细胞间病毒
鼠伤寒沙门氏菌	能引起肠胃炎的细胞性病原体 在水中代谢旺盛

附录 3 确定 MAR 砷和铁释放潜力的决策树

要理解 MAR 地球化学过程中危害物释放的潜力，需要下面的知识：

- 原水水质；
- 本地地下水水质；
- 含水层基质组成；
- 矿物学；
- 主要和次要因素；
- 变换潜力；
- 有机碳含量。

附图 1 和附图 2 给出了第 4 部分所讨论的确定 MAR 方案中砷和铁释放潜力的决策树。储存阶段沉积物中砷的空间扩散变化极大，很难预测砷的释放潜力。决策树不是地球化学模型，但可以用于帮助理解重要的地球化学过程。

砷释放同时可以指示其他危害物的释放，例如：钴、铬、铜、镍和锌。地下水中的砷浓度可以用来指示沉积物中的砷，报告称其检测下限为 1 μg/L。提高检测下限(如<5 μg/L)不能防止砷的释放量超过指标值。

铁的决策树可用于确定铁释放潜力和后续恢复水的铁增加，或者反映补给点附近由铁的氧化物的降落造成的含水层堵塞。因为锰的表面性质与铁类似，所以如果含水层的主要成分是锰，可以借用铁的决策树来检测锰释放的潜力。

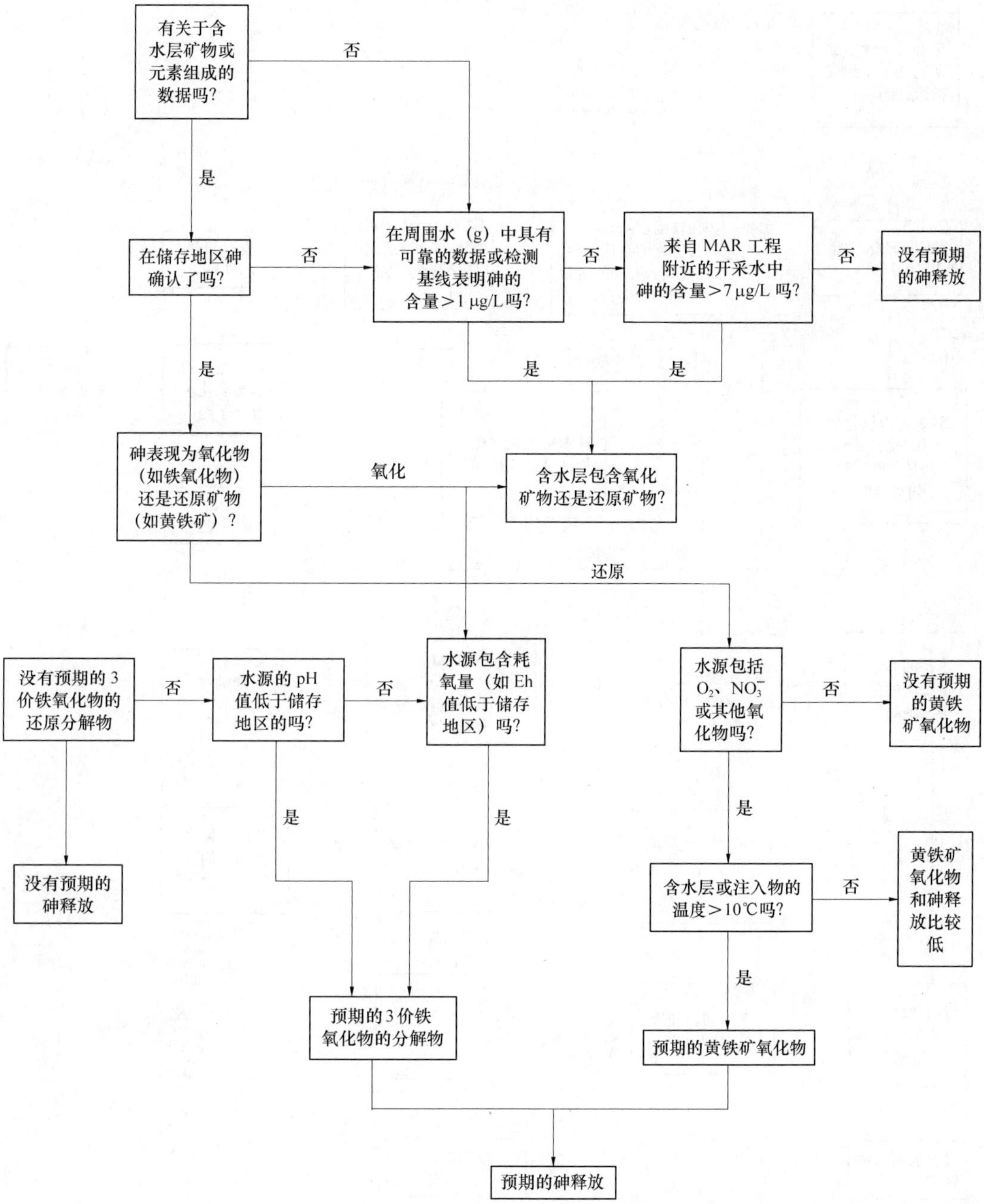

附图 1　在含水层储存与补给中关于识别砷释放的决策树，砷浓度应该是 1 μg/L，其他痕量金属也同时释放

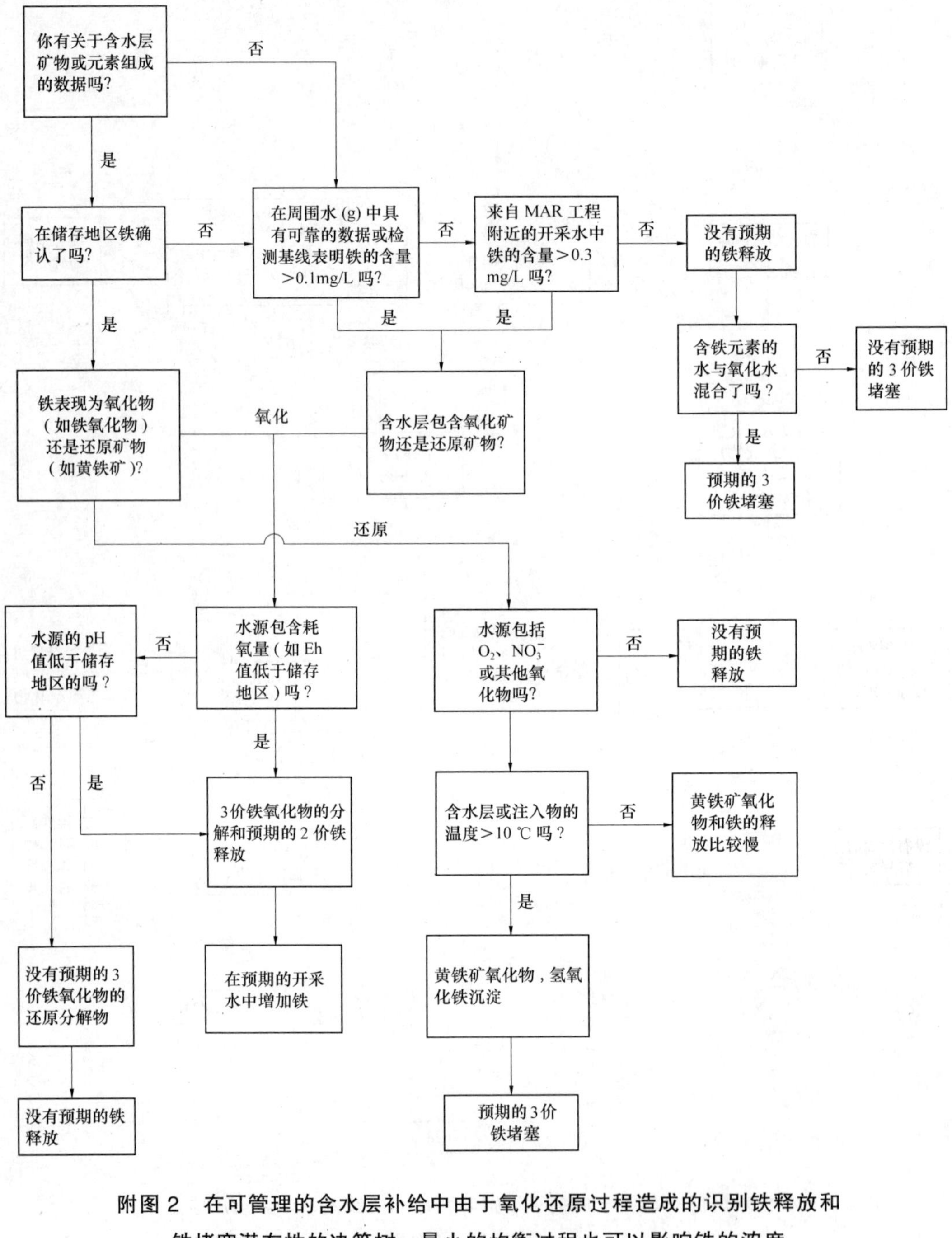

附图2　在可管理的含水层补给中由于氧化还原过程造成的识别铁释放和铁堵塞潜在性的决策树；最小的均衡过程也可以影响铁的浓度